S0-BOF-993

COAL
RESEARCH AND DEVELOPMENT
TO SUPPORT NATIONAL ENERGY POLICY

Committee on Coal Research, Technology, and
Resource Assessments to Inform Energy Policy

Board on Earth Sciences and Resources

Division on Earth and Life Studies

NATIONAL RESEARCH COUNCIL
OF THE NATIONAL ACADEMIES

THE NATIONAL ACADEMIES PRESS
Washington, D.C.
www.nap.edu

THE NATIONAL ACADEMIES PRESS   **500 Fifth Street, N.W.**   **Washington, DC 20001**

NOTICE: The project that is the subject of this report was approved by the Governing Board of the National Research Council, whose members are drawn from the councils of the National Academy of Sciences, the National Academy of Engineering, and the Institute of Medicine. The members of the committee responsible for the report were chosen for their special competences and with regard for appropriate balance.

The opinions, findings, and conclusions or recommendations contained in this document are those of the authors and do not necessarily reflect the views of the Office of Surface Mining Reclamation and Enforcement. Mention of trade names or commercial products does not constitute their endorsement by the U.S. government. Supported by the Office of Surface Mining Reclamation and Enforcement, Department of the Interior, under Award No. CT5-06401.

International Standard Book Number 13:   978-0-309-11022-8 (Book)
International Standard Book Number 10:   0-309-11022-X (Book)
International Standard Book Number 13:   978-0-309-11023-5 (PDF)
International Standard Book Number 10:   0-309-11023-8 (PDF)

Library of Congress Control Number: 200793585

Additional copies of this report are available from the National Academies Press, 500 Fifth Street, N.W., Lockbox 285, Washington, DC 20055; (800) 624-6242 or (202) 334-3313 (in the Washington metropolitan area); Internet http://www.nap.edu

Cover: Cover design by Michele de la Menardiere; upper image courtesy of Timothy J. Rohrbacher, USGS; lower image courtesy CONSOL Energy Inc.

Copyright 2007 by the National Academy of Sciences. All rights reserved.

Printed in the United States of America.

THE NATIONAL ACADEMIES
Advisers to the Nation on Science, Engineering, and Medicine

The **National Academy of Sciences** is a private, nonprofit, self-perpetuating society of distinguished scholars engaged in scientific and engineering research, dedicated to the furtherance of science and technology and to their use for the general welfare. Upon the authority of the charter granted to it by the Congress in 1863, the Academy has a mandate that requires it to advise the federal government on scientific and technical matters. Dr. Ralph J. Cicerone is president of the National Academy of Sciences.

The **National Academy of Engineering** was established in 1964, under the charter of the National Academy of Sciences, as a parallel organization of outstanding engineers. It is autonomous in its administration and in the selection of its members, sharing with the National Academy of Sciences the responsibility for advising the federal government. The National Academy of Engineering also sponsors engineering programs aimed at meeting national needs, encourages education and research, and recognizes the superior achievements of engineers. Dr. Charles M. Vest is president of the National Academy of Engineering.

The **Institute of Medicine** was established in 1970 by the National Academy of Sciences to secure the services of eminent members of appropriate professions in the examination of policy matters pertaining to the health of the public. The Institute acts under the responsibility given to the National Academy of Sciences by its congressional charter to be an adviser to the federal government and, upon its own initiative, to identify issues of medical care, research, and education. Dr. Harvey V. Fineberg is president of the Institute of Medicine.

The **National Research Council** was organized by the National Academy of Sciences in 1916 to associate the broad community of science and technology with the Academy's purposes of furthering knowledge and advising the federal government. Functioning in accordance with general policies determined by the Academy, the Council has become the principal operating agency of both the National Academy of Sciences and the National Academy of Engineering in providing services to the government, the public, and the scientific and engineering communities. The Council is administered jointly by both Academies and the Institute of Medicine. Dr. Ralph J. Cicerone and Dr. Charles M. Vest are chair and vice chair, respectively, of the National Research Council.

www.national-academies.org

COMMITTEE ON COAL RESEARCH, TECHNOLOGY, AND RESOURCE ASSESSMENTS TO INFORM ENERGY POLICY

CORALE L. BRIERLEY, *Chair*, Brierley Consultancy LLC, Highlands Ranch, Colorado
FRANCIS P. BURKE, CONSOL Energy Inc. *(retired)*, South Park, Pennsylvania
JAMES C. COBB, University of Kentucky, Lexington
ROBERT B. FINKELMAN, University of Texas at Dallas
WILLIAM FULKERSON, Institute for a Secure and Sustainable Environment, University of Tennessee, Knoxville
HAROLD J. GLUSKOTER, U.S. Geological Survey *(emeritus)*, McLean, Virginia
MICHAEL E. KARMIS, Virginia Polytechnic Institute and State University, Blacksburg
KLAUS S. LACKNER, Columbia University, New York
REGINALD E. MITCHELL, Stanford University, California
RAJA V. RAMANI, The Pennsylvania State University, University Park
JEAN-MICHEL M. RENDU, Mining Consultant, Englewood, Colorado
EDWARD S. RUBIN, Carnegie Mellon University, Pittsburgh, Pennsylvania
SAMUEL A. WOLFE, New Jersey Board of Public Utilities, Newark

National Research Council Staff
DAVID A. FEARY, Study Director
TANYA PILZAK, Research Associate (until December 2005)
CAETLIN M. OFIESH, Research Associate (January-March 2006)
KRISTEN B. DALY, Research Associate (March-July 2006)
SANDI SCHWARTZ, Project Researcher (from August 2006)
JENNIFER T. ESTEP, Financial and Administrative Associate
JAMES DAVIS, Senior Project Assistant (until December 2005)
AMANDA M. ROBERTS, Senior Project Assistant (January-August 2006)
NICHOLAS D. ROGERS, Senior Project Assistant (from August 2006)

COMMITTEE ON EARTH RESOURCES

MURRAY W. HITZMAN, *Chair,* Colorado School of Mines, Golden
FRANCIS P. BURKE, CONSOL Energy Inc. *(retired)*, South Park,
  Pennsylvania
WILLIAM S. CONDIT, Independent Consultant, Santa Fe, New Mexico
MICHAEL DOGGETT, Queen's University, Kingston, Ontario, Canada
THOMAS V. FALKIE, Berwind Natural Resources Corporation *(retired)*,
  Newtown Square, Pennsylvania
PATRICIA M. HALL, BP America Inc., Houston, Texas
DAVID D. LAURISKI, Safety Solutions International, LLC, Parker, Colorado
ANN S. MAEST, Stratus Consulting, Boulder, Colorado
LELAND L. MINK, U.S. Department of Energy Geothermal Program *(retired)*,
  Worley, Idaho
REGINAL SPILLER, Frontera Resources Corporation, Houston, Texas
SAMUEL J. TRAINA, University of California, Merced
HAROLD J. VINEGAR, Shell Exploration and Production Company, Houston,
  Texas

National Research Council Staff
ELIZABETH A. EIDE, Senior Program Officer
NICHOLAS D. ROGERS, Senior Program Assistant

BOARD ON EARTH SCIENCES AND RESOURCES

GEORGE M. HORNBERGER, *Chair*, University of Virginia, Charlottesville
GREGORY B. BAECHER, University of Maryland, College Park
STEVEN R. BOHLEN, Joint Oceanographic Institutions, Washington, D.C.
KEITH C. CLARKE, University of California, Santa Barbara
DAVID COWEN, University of South Carolina, Columbia
WILLIAM E. DIETRICH, University of California, Berkeley
ROGER M. DOWNS, The Pennsylvania State University, University Park
JEFF DOZIER, University of California, Santa Barbara
KATHERINE H. FREEMAN, The Pennsylvania State University, University Park
RHEA L. GRAHAM, Pueblo of Sandia, Bernalillo, New Mexico
RUSSELL J. HEMLEY, Carnegie Institute of Washington, Washington, D.C.
MURRAY W. HITZMAN, Colorado School of Mines, Golden
LOUISE H. KELLOGG, University of California, Davis
V. RAMA MURTHY, University of Minnesota, Minneapolis
CLAYTON NICHOLS, Idaho National Engineering and Environmental
     Laboratory *(retired)*, Sandpoint
RAYMOND A. PRICE, Queen's University, Ontario, Canada
BARBARA A. ROMANOWICZ, University of California, Berkeley
JOAQUIN RUIZ, University of Arizona, Tucson
MARK SCHAEFER, Global Environment and Technology Foundation,
     Arlington, Virginia
WILLIAM W. SHILTS, Illinois State Geological Survey, Champaign
RUSSELL STANDS-OVER-BULL, BP American Production Company,
     Houston, Texas
TERRY C. WALLACE, JR., Los Alamos National Laboratory, New Mexico
THOMAS J. WILBANKS, Oak Ridge National Laboratory, Tennessee

National Research Council Staff
ANTHONY R. DE SOUZA, Director
PAUL M. CUTLER, Senior Program Officer
ELIZABETH A. EIDE, Senior Program Officer
DAVID A. FEARY, Senior Program Officer
ANNE M. LINN, Senior Program Officer
ANN G. FRAZIER, Program Officer
SAMMANTHA L. MAGSINO, Program Officer
CAETLIN M. OFIESH, Associate Program Officer
RONALD F. ABLER, Senior Scholar
JENNIFER T. ESTEP, Financial and Administrative Associate
VERNA J. BOWEN, Financial and Administrative Associate
JARED P. ENO, Research Associate
NICHOLAS D. ROGERS, Research Associate
TONYA E. FONG YEE, Program Assistant

vi

Preface

The extraordinarily broad scope of the congressional request for advice on coal resources and future coal research and development needs provided a significant challenge for the committee appointed by the National Research Council (NRC). Fortunately, clarifications by staff members from the offices of U.S. Senators Robert C. Byrd and Arlen Specter—the originators of this study—were most helpful, suggesting that the report should be brief and contain limited detail, but with abundant references to other, more comprehensive studies. They also emphasized that a major element of their request was to learn of any potential roadblocks that might impinge on the production or delivery of coal should the nation's energy requirements dictate that a substantial increase in coal use was needed.

The task for the committee was made easier by the many experts in all aspects of the coal life cycle who freely gave up their time to make presentations in open session. These presentations formed the basis for the committee's deliberations as it fashioned the findings and recommendations. The committee's task was also facilitated by the cooperation of the interagency liaison group, established and coordinated by the Office of Surface Mining Reclamation and Enforcement (OSM), which provided input to the committee at its public meetings and responded to specific questions.

I am truly indebted to the committee members, all of whom remained completely engaged in the entire process from start to finish. All gave generously of their expertise, time, and energy, and provided wit and cheerfulness when they were sorely needed. Collectively, they performed as a skillful team with dedication and determination. On behalf of the committee I thank the NRC staff: David Feary, whose input and guidance was indispensable in producing a focused and lucid report; Anthony de Souza, Tanya Pilzak, Caetlin Ofiesh, Kristen Daly, and Sandi Schwartz, who assisted with broad guidance and background information; and James Davis, Amanda Roberts, and Nicholas Rogers, who made sure the committee process proceeded efficiently and effectively.

Corale L. Brierley
Chair

Acknowledgments

This report was greatly enhanced by input from the many participants at the public committee meetings held as part of this study—Mike Adamczyk, Carl O. Bauer, Peter J. Bethell, Perry Bissell, Paul Bollinger, Richard Bonskowski, Wanda Burget, Gregory E. Conrad, John Craynon, Rob Donovan, Tom Dower, Mike Eastman, Nick Fedorko, Sara Flitner, Bradford Frisby, Ari Geertsema, Steve Gigliotti, Thomas J. Grahame, Güner Gürtunca, David Hawkins, Peter Holman, Connie Holmes, Mike Hood, James R. Katzer, Larry Kellerman, Julianne M. Klara, Mo Klefeker, Jeffrey L. Kohler, John Langton, John A. Lewis, Alexander Livnat, James Luppens, Gerald H. Luttrell, Maria M. Mitchell, John Moran, M. Granger Morgan, Mike Mosser, John Novak, Karen Obenshain, Bruce Peterson, Brenda S. Pierce, Jacek Podkanski, Craig Rockey, Timothy Rohrbacher, Scott Sitzer, Neil Stiber, Eugene Trisko, Ted Venners, Kimery Vories, Franz Wuerfmannsdobler, and Ben Yamagata. These presentations and the ensuing discussions helped set the stage for the committee's fruitful discussions in the sessions that followed. We also gratefully acknowledge the people who facilitated our committee meetings, the company personnel who briefed the committee on mine operations and led the committee on mine and plant tours, and the experts who supplied information in response to specific enquiries by the committee—David Aloia, Gene D. Berry, Joe Cerenzia, Becki Dale, Mark Davies, James Dooley, Bob Green, Mark Kamlet, Gary G. Loop, James Manual, Claudia L. Miller, Phillip H. Nicks, Jack C. Pashin, Mark Payne, Joe Vaccari, Marshall Wise, and Connie Zaremsky.

This report has been reviewed in draft form by individuals chosen for their diverse perspectives and technical expertise, in accordance with procedures approved by the National Research Council's Report Review Committee. The purpose of this independent review is to provide candid and critical comments that will assist the institution in making its published report as sound as possible and to ensure that the report meets institutional standards for objectivity, evidence, and responsiveness to the study charge. The review comments and draft manuscript remain confidential to protect the integrity of the deliberative process. We wish to thank the following individuals for their participation in the review of this report:

Heinz H. Damberger, Illinois State Geological Survey (retired), Boulder, Colorado

Mark Davies, Rio Tinto Energy America, Gillette, Wyoming

Thomas V. Falkie, Berwind Natural Resources Corporation (retired), Newtown Square, Pennsylvania

Barbara A. Filas, Knight Piesold and Company, Denver, Colorado

Paul E. Gray, Department of Electrical Engineering and Computer Science, Massachusetts Institute of Technology, Cambridge

R. Larry Grayson, School of Mining and Metallurgy, University of Missouri, Rolla

Howard J. Herzog, Laboratory for Energy and the Environment, Massachusetts Institute of Technology, Cambridge

John N. Murphy, Department of Chemical and Petroleum Engineering, University of Pittsburgh, Pennsylvania

Dianne R. Nielson, Utah Department of Environmental Quality, Salt Lake City

Lee Saperstein, School of Mines and Metallurgy, University of Missouri-Rolla (retired), Nantucket, Massachusetts

Stanley C. Suboleski, Federal Mine Safety and Health Review Commission (retired), Midlothian, Virginia

Although the reviewers listed above provided many constructive comments and suggestions, they were not asked to endorse the conclusions or recommendations nor did they see the final draft of the report before its release. The review of this report was overseen by William G. Agnew, General Motors Corporation (retired), Corrales, New Mexico, and William L. Fisher, Jackson School of Geosciences, the University of Texas, Austin. Appointed by the National Research Council, they were responsible for making certain that an independent examination of this report was carried out in accordance with institutional procedures and that all review comments were carefully considered. Responsibility for the final content of this report rests entirely with the authoring committee and the institution.

Contents

Summary

T he coal industry in the United States—encompassing coal mining, process-ing, and transportation—is a relatively small but vitally important compo-nent of the nation's economy. Coal provides nearly a quarter of all energy supplies in the United States, with most of this coal used to generate more than half of the nation's electricity. The expectation of continually increasing national electricity demand has led to forecasts suggesting that the demand for coal may increase by 60 to 70 percent over the next 25 years, although other analyses suggest that coal use may grow at a slower rate—or even decline—depending on the timing and magnitude of regulatory limits on carbon dioxide emissions. With this degree of uncertainty, coal-related research and development (R&D) policies need to accommodate a broad range of possible future scenarios. Con-gress asked the National Research Council to undertake a broad examination of coal-related R&D across the entire fuel cycle (see Box S.1), with briefings by congressional staff emphasizing that the study should be brief, should concentrate on the "upstream"[1] aspects of the coal industry and deal only briefly with coal utilization R&D, and should highlight any potential stumbling blocks to increased coal production.

The context for any assessment of future coal production is inextricably linked with the development of a national carbon emissions policy. Potential constraints on greenhouse gas (especially CO_2) emissions, and the technical and economic feasibility of CO_2 control measures, are the dominant issues affecting the outlook for the future of coal use over the next 25 years and beyond. The

---

[1] Upstream activities refer to pre-utilization processes—coal mining, processing, and transport to utilization sites.

BOX S.1
Statement of Task

The study will broadly examine coal resource assessments, technologies, and research and development (R&D) activities in the United States in order to formulate an appropriate, integrated roadmap of future needs. The results of the review should help define and construct a national strategy for coal R&D and resource assessments.

The study shall consider the following issues:

1. Summarize recent projections of the coal use as part of the U.S. and global energy portfolios over the next 25 years, including projections that take into account the potential roles of coal in future integrated energy and environmental policies, in order to set the context for development of a more comprehensive, strategic roadmap for coal R&D and resource assessments.

2. Describe the full range of local, regional, national, and global issues and challenges, including environmental issues that must be taken into account when considering future production and utilization of coal.

3. Review the coal reserve assessments based on recent trends in the coal sector and examine the current and future role of coal imports and exports.

4. Assess the categories of coal R&D currently being carried out in the United States and internationally, and investigate whether and how technology developments in other fields can be applied to the coal sector. Review how technologies are being transferred to coal mine operators and other users, recognizing differences among companies.

5. Determine the priority coal R&D needs, including in the areas of exploration, discovery, reserve assessment (including in terms of commercial feasibility for known reserves), extraction, coal preparation, delivery to market, waste disposal, reclamation, health and safety, community impact, environmental practices, education and training, and productivity.

6. Evaluate the need for a broad-based, coordinated, multi-agency coal research and development program. Review current coal-related research, examine what agencies are conducting it, and determine how much funding is currently being spent throughout the coal life cycle.

7. Examine options for supporting and implementing a broad-based coal R&D program, including approximate costs, and the relative roles and commitments of the public and private sectors now and into the future.

---

difficulty of predicting the prices and availability of alternative energy sources for electric power generation provides additional uncertainty. Taking these factors into consideration, an assessment of forecasts for coal use indicates that over the next 10 to 15 years (until about 2020), coal production and use in the United States are projected to range from about 25 percent above to about 15 percent below 2004 levels, depending on economic conditions and environmental policies. By 2030, the range of projected coal energy use in the United States broadens considerably, from about 70 percent above to 50 percent below current levels.

The higher values reflect scenarios with high oil and gas prices and no restrictions on carbon emissions. The lower values reflect scenarios with relatively strict limits on U.S. CO_2 emissions, which cause coal use with sequestration to be more costly compared to other options for power generation.

At present, coal imports and exports represent small fractions of total U.S. coal production and use, and projections indicate that both imports and exports are expected to remain relatively small. From a global perspective, the largest tonnage increases in coal use are expected in the emerging economies of China and India. Much smaller tonnage growth is projected in the rest of the world, although relative growth rates are projected to be high in several other countries. Again, however, there is great uncertainty in projections of global coal use, especially beyond about 2020.

These projections provide the context for an assessment of coal-related R&D activities. A number of organizations and entities—federal government agencies, state government agencies, academic institutions, coal mining companies, and equipment manufacturers—are engaged in aspects of coal-related R&D and technology development. In this report, the primary focus is on federal government support for activities that are variously described as pure research, applied science, pilot-scale testing, technical support, demonstration projects, and applied engineering projects. For existing federal support, the committee analyzed R&D budgets in terms of the range of categories that encompass the coal fuel cycle—resource and reserve assessment; coal mining and processing; coal mining safety and health; environmental protection and reclamation; transport of coal and coal-derived products; and coal utilization.

There are numerous applied research areas, focused primarily on incremental technology development, for which federal involvement is neither appropriate nor required and where industry should and does provide support. For some areas, such as ensuring that a well-trained workforce is available to meet the nation's mining and mining education requirements, federal involvement can effectively complement industry activities. There are other areas of coal-related R&D in which the federal government has a primary role—for example, to establish the quantity and quality of the nation's coal reserves, to facilitate and catalyze revolutionary (rather than incremental) technology development, to safeguard the health and safety of mine workers, and to protect the environment during future mining and processing and mitigate existing environmental problems arising from past mining practices. It is also a federal responsibility to provide funding for the R&D required to support the government's regulatory role.

More than $538 million was spent by federal government agencies for coal-related research and technology development in 2005. Of this, more than 90 percent (~$492 million) was directed towards "downstream" aspects of coal use, mostly coal utilization technology development and transmission research funded through the U.S. Department of Energy (DOE). Federal support for R&D activities related to all upstream aspects of the coal fuel cycle (i.e., mine worker safety and health, resource and reserve assessments, coal mining and processing, and environmental protection and reclamation) accounted for less than 10 percent of the total federal investment in coal-related R&D. Federal funding in 2005 for individual components of upstream activities ranged from $24.4 million (4.5 percent) for mine worker safety and health R&D to $1.3 million (0.2 percent) for coal mining and processing R&D.

Consideration of agency budgets over the past 10 to 15 years shows that federal government funding of R&D to support its regulatory role has remained broadly constant. In contrast, support for coal resource and reserve assessments has declined by nearly 30 percent as inflation has eroded constant nominal dollar funding, and support for mining and processing research declined dramatically in the mid-1990s, coinciding with the dissolution of the U.S. Bureau of Mines, and now represents only 0.2 percent of total federal coal-related R&D funding.

There are some components of the coal fuel cycle (e.g., coal transportation) where identification of potential stumbling blocks that may impede increased coal production and use do not lead to R&D recommendations—these issues are more appropriately dealt with by regulatory actions and existing government authority or will ultimately be resolved by standard business practices. However, for most components of the coal fuel cycle, a range of national interests—the need for sound information on which to base policy decisions, the requirement for optimum use of an important national resource, or society's demand for personal or environmental health and safety—lead to a series of recommendations for high-priority R&D activities; these are noted below in bold.

COAL RESOURCE, RESERVE, AND QUALITY ASSESSMENTS

Federal policy makers require accurate and complete estimates of national coal reserves to formulate coherent national energy policies. Despite significant uncertainties in existing reserve estimates, it is clear that there is sufficient coal at current rates of production to meet anticipated needs through 2030. Further into the future, there is probably sufficient coal to meet the nation's needs for more than 100 years at current rates of consumption. However, it is not possible to confirm the often-quoted assertion that there is a sufficient supply of coal for the next 250 years. A combination of increased rates of production with more detailed reserve analyses that take into account location, quality, recoverability, and transportation issues may substantially reduce the number of years of supply. Future policy will continue to be developed in the absence of accurate estimates until more detailed reserve analyses—which take into account the full suite of geographical, geological, economic, legal, and environmental characteristics—are completed.

Present estimates of coal reserves are based upon methods that have not been reviewed or revised since their inception in 1974, and many of the input data were compiled in the early 1970s. Recent programs to assess reserves in limited areas using updated methods indicate that only a small fraction of previously estimated reserves are economically recoverable. Such findings emphasize the need for a reinvigorated coal reserve assessment program using modern methods and technologies to provide a sound basis for informed decision making.

> **A coordinated federal-state-industry initiative to determine the magnitude and characteristics of the nation's recoverable coal reserves, using modern mapping, coal characterization, and database technologies, should be instituted with the goal of providing policy makers with a comprehensive accounting of national coal reserves within 10 years.**

The committee recommends that the U.S. Geological Survey should lead a federal-state-industry initiative to quantify and characterize the nation's coal reserves, and estimates that this will require additional funding of approximately $10 million per year.

RESEARCH TO SUPPORT COAL MINING AND PROCESSING

Regardless of the precise levels of future coal production, the coal mines of the future will encounter a range of new or more difficult mining and processing challenges as more easily accessed coal seams are depleted and the industry turns to less accessible reserves. Surface operations will mine deeper seams that require increased stripping ratios and multiple benches, and underground mines will need to access seams that are deeper, thinner, or thicker, generally with higher methane content and potentially presenting greater difficulties with strata control. These more difficult mining conditions will require improved methods to protect the health and safety of mine workers, careful environmental management of mined lands and waste products, and improved recovery to optimize use of the nation's coal resource.

Improved Mine Worker Health and Safety

A range of factors increase health and safety risks to the coal mining workforce, including the introduction of new equipment and systems; the commencement of mining in virgin areas; the infusion of new workers; and the mining of multiple seams and seams that are thinner, thicker, or deeper than those customarily mined at present, as well as new seams that underlie or overlie previously mined-out seams. All of these factors are likely to apply to some degree in future mines, and such risks are likely to become more pronounced if coal production levels increase. There are major knowledge gaps and technology needs in the areas of survival, escape, communications systems (both surface-to-underground and underground-to-underground), and emergency preparedness and rescue. Additional risk factors that are likely to apply in the deeper mines of the future are the potential hazards related to methane control, dust control, ignition sources, fires, and explosions. A greater understanding and better prediction of strata control to prevent unanticipated roof collapse are essential for maintaining and improving worker safety.

Health and safety research and development should be expanded to anticipate increased hazards in future coal mines. These R&D efforts should emphasize improved methane control, improved mine ventilation, improved roof control, reduced repetitive and traumatic injuries, reduced respiratory diseases, improved escape and rescue procedures, improved communications systems, and research to reduce explosions and fires. This should be coupled with improved training of the mining workforce in all aspects of mine safety. R&D should also be directed toward lowering the exposure of mine workers to hazardous conditions, particularly through expanded use of remote sensing and the automation of mining operations.

Most mining health and safety research by the federal government is carried out by the Mining Program at the National Institute for Occupational Safety and Health (NIOSH). Technology-related activities within the Mine Safety and Health Administration (MSHA) are limited to technical support and training services for its personnel and those from the mining industry. With NIOSH carrying out the research needed to improve mine safety and support MSHA's regulatory role, these two agencies play a vital role in coal mine worker health and safety. The committee estimates that the enhanced health and safety program proposed here will require additional annual R&D funding of approximately $35 million, and recommends that NIOSH continue as the lead agency with enhanced coordination with MSHA and industry.

Improved Environmental Protection

As mining extracts coal from deeper and operationally more difficult seams by both surface and underground methods, a range of existing environmental issues and concerns will be exacerbated, and new concerns—particularly related to greater disturbance of hydrologic systems, ground subsidence, and waste management at mines and preparation plants—are likely to arise. Inadequate understanding of post-mining behavior of strata, stability of spoils, and the associated hydrologic consequences of mining in both surface and underground mines affects mine permitting, mine development, environmental mitigation, and post-mining land use, including use for waste management. Research offers considerable potential to mitigate the effects of past mining practices, particularly acid mine drainage on abandoned mine lands. However, the regulatory environment and the technical support programs administered by both state and federal agencies, and implemented by mining companies through their compliance practices, are inadequately supported by existing research programs.

Additional research is needed to mitigate the adverse environmental impacts associated with past, existing, and future coal mining and processing. Research activities should focus particularly on developing techniques to mitigate the alteration and collapse of strata overlying mined areas, to model the hydrological impacts of coal mining, to improve mine mapping and void detection, to improve the stability of spoils on steep slopes, and to improve the construction and monitoring of impoundments.

Both the Office of Surface Mining Reclamation and Enforcement (OSM) and the U.S. Environmental Protection Agency (EPA), although primarily regulatory agencies, fund limited R&D activities in support of their missions. The committee estimates that additional funding of approximately $60 million per year will be required to conduct the research necessary to adequately respond to the environmental impacts of past, existing, and future mining operations. The committee recommends that OSM should be the lead agency in this effort, and it should coordinate closely with related EPA and state research activities.

Improved Mine Productivity and Resource Optimization

Although technology developments (primarily underground longwall mining) and industry changes (primarily the growth in large surface operations) resulted in a two- to three-fold increase in the productivity of U.S. coal mines over the past three decades, production and productivity increases in recent years have been small as mining companies and equipment manufacturers have made only incremental improvements. Over the past decade, there has been little R&D directed toward truly advanced mining technologies, and at present, only 0.2 percent of total federal coal-related R&D funding is directed toward development of the advanced mining technologies and practices that are necessary to optimize utilization of the nation's coal resource. Small percentage increases in coal recovery through improved mining and coal preparation processes have the potential to significantly expand economically recoverable reserves of both eastern and western coals. The development of these technologies, increasingly needed as coal reserve quality decreases over time, will help to maximize utilization of the nation's coal resource.

The global transfer of coal mining and processing technology within the industry is facilitated by international equipment manufacturers, who work closely with suppliers and the larger mining clients on evolutionary product developments. However, there is little evidence of the efficient transfer of technologies from outside the mining industry. This is at least partly due to the relatively small market that the coal mining industry represents to potential technology suppliers and the scarcity of coal mining research at academic institutions and national laboratories.

There should be renewed support for advanced coal mining and processing research and development to optimize use of the nation's coal resources by increasing the amount of coal that is economically minable through technological advances that accommodate health, safety, and environmental requirements. The focus of this R&D should be on increased integration of modern technology in the extraction and processing phases of coal production, with particular emphasis on emerging advances in materials, sensors, and controls; monitoring; and automated mining systems.

Research to develop advanced mining technologies requires not only cooperation among relevant federal agencies, but also participation by academic institutions as well as funding, guidance, and technology transfer by industry. The committee estimates that advanced coal mining and processing R&D will require a total of approximately $60 million per year and recommends that this funding should comprise $30 million in total federal support, with cost sharing from non-federal sources. The DOE Office of Fossil Energy (DOE-FE) should be the lead federal agency and should coordinate with the National Science Foundation (NSF), OSM, NIOSH, academic institutions, and the coal industry to ensure that all research activities carefully consider the environmental, reclamation, and health and safety aspects of coal mining.

TRANSPORT OF COAL AND COAL PRODUCTS

Growth in the use of coal depends on having sufficient capacity to deliver increasing amounts of coal reliably and at reasonable prices to an end user. The capacity, reliability, and price of rail transportation—the dominant mode of coal transport—depend largely on the supply and demand for rail transportation, as well as on prevailing business practices, the investment climate, and the nature of regulatory oversight of the railroad industry. The capacity, reliability, and price of rail transportation of coal depend to a far lesser degree upon research and development. Reliable and sufficient waterborne transportation—the second most prevalent method of coal transport—depends on the construction and maintenance of waterway infrastructures, especially lock-and-dam infrastructure and port capacity.

Much of the nation's coal-fired electric generating capacity is located at some distance from the urbanized areas that have the largest and most concentrated demands for electricity. Projections of higher coal use depend on sufficient capacity to transmit electricity from coal-based power plants to such areas reliably and at a reasonable cost. Conversely, the projected increases in coal use will diminish if these high-demand areas satisfy much of their growing demand for electricity not by expanding their ability to import electricity from areas where coal is plentiful, but by a combination of energy efficiency, demand response, and local electric generation from sources other than coal.

The coal transportation and electric power transmission systems are large and complex networks in which localized disruptions can have severe and widespread impacts. Weather and other natural phenomena, as well as societal factors such as sabotage and terrorism, impose a range of risks on these systems. These characteristics make it difficult to guarantee that there will be sufficient capacity to transport coal or coal-based energy (primarily electricity) reliably and cost-effectively to the various end users, particularly in light of scenarios that predict substantially increased coal use. Research is needed to better understand the factors that control these large and complex networks to minimize the risks of cascading system disruptions.

RESEARCH TO SUPPORT COAL UTILIZATION

In accord with requests that this study focus primarily on the upstream aspects of the coal fuel cycle, the analysis of coal utilization R&D is confined to a brief overview that is primarily focused on describing the factors associated with coal use that are most likely to impose constraints on future demands for coal. Overwhelmingly, the environmental impacts of coal use, especially carbon dioxide emissions associated with global climate change, pose the greatest potential constraint on future coal utilization. Decisions to invest or not invest in coal-based power plants will strongly influence future coal use, and these decisions will depend in large part on the timing and magnitude of any future constraints on CO_2 emissions.

In contrast, potential regulatory requirements to further reduce emissions of NO_x, SO_2, mercury, and particulate matter in the future are not expected to significantly limit the overall use of coal in the next several decades. However, future emission control requirements for these regulated air pollutants could result in changed preferences for particular types of coal, depending on the nature of future regulations.

If coal is to continue as a primary component of the nation's future energy supply in a carbon-constrained world, large-scale demonstrations of carbon management technologies—especially carbon capture and sequestration (CCS)—are needed to prove the commercial readiness of technologies to significantly reduce CO_2 emissions from coal-based power plants and other energy conversion processes. In addition, detailed assessments are needed to identify potential geological formations in the United States that are capable of sequestering large quantities of CO_2; to quantify their storage capacities; to assess migration and leakage rates; and to understand the economic, legal, and environmental impacts of storage on both near-term and long-term time scales. These R&D activities would complement other legal and regulatory activities needed to make these sites available and viable as a CO_2 control strategy. Such geologic sequestration sites should be considered "resources," and categorized and described in the same way that conventional mineral or energy resources are assessed.

The U.S. Geological Survey (USGS) should play a lead role in identifying, characterizing, and cataloguing the CO_2 sequestration capacity of potential geologic sequestration resources.

The committee estimates that approximately $10 million per year for five years will be required for this activity, which would be in addition to the CCS research and demonstration program presently under way at DOE. There should be close cooperation and coordination among the USGS, the Carbon Sequestration Program managed by DOE's Office of Fossil Energy, and the states involved in DOE's Regional Carbon Sequestration Partnerships.

COORDINATION OF COAL-RELATED R&D BY FEDERAL AGENCIES

One component of this study was the specific requirement for the committee to evaluate whether a broad-based, coordinated, multiagency coal R&D program is required, and if so, to examine options for supporting and implementing such a program. The committee carefully considered existing R&D programs and assessed the extent of—and opportunities for—coordination of coal-related research among the agencies. The committee also considered coal-related R&D support provided by states, the coal industry, and equipment manufacturers, but did not attempt an exhaustive compilation of these non-federal activities. The committee concluded that rather than proposing a single "mega-agency," improved interagency coordination to respond to specific R&D opportunities and challenges could be better implemented through cooperation among two or more federal entities in R&D partnerships, with involvement of non-federal bodies as appropriate. A number of key factors contributed to this conclusion—the highly varied mandates of the various agencies or offices, in some cases with specific single-focus missions (e.g., regulatory role of MSHA, basic research role of NSF, applied research role of NIOSH), whereas other agencies or offices have broader mandates (e.g., EPA's regulatory and R&D roles, DOE's wide-ranging mission that also includes support for demonstration projects); their capacities for conducting or managing R&D programs; and the different congressional committees that have responsibility for their funding and oversight.

Accordingly, much stronger R&D partnerships should be established in the areas of coal resource and reserve assessment (**USGS**,[2] DOE-EIA, states, industry); improved mine worker health and safety (**NIOSH**, MSHA, indus-

---

[2]Recommended lead agencies are shown in bold.

try); improved environmental protection (**OSM**, EPA, states, industry); improved resource recovery and mine productivity (**DOE-FE**, NSF, OSM, NIOSH, academic institutions, industry); and carbon sequestration resource characterization (**USGS**, DOE-FE, states). The total new funding to support these activities amounts to approximately $144 million per year (Table S.1).

SOCIETAL ISSUES

While coal mining benefits communities during the productive life of a mine, after mine closure there is the potential for adverse affects that may include land use, safety, infrastructure and community development, and sustainability issues. The key to maintaining healthy communities after cessation of mining is early and comprehensive planning that involves all stakeholders.

An aging workforce and a substantial shortage of technically trained personnel in the mining and minerals engineering disciplines pose a threat to projected scenarios that involve substantially increased coal production. Extramural funding by federal agencies to universities in support of research in earth sciences and engineering would assist in recruiting, retaining, and developing mining professionals. This extramural funding is expected to be supported by proposed increased funding to the federal agencies summarized in Table S.1.

TABLE S.1 Summary of FY 2005 and Proposed Additional Funding for Coal-Related R&D at Federal Agencies

	FY 2005 Funding (million dollars)	Proposed New Funding (million dollars)	Total Proposed Funding (million dollars)
Resource and reserve assessments and characterization	10[a]	20[a]	30[a]
Improved mine worker health and safety	25	35	60
Environmental protection and reclamation	10	60	70
Improved mining productivity and resource optimization	1	29	30
Total	**46**	**144**	**190**

NOTE: All figures are in millions of dollars per year. FY 2005 figures are rounded to nearest million for easier comparison with proposed funding levels (unrounded figures for FY 2005 funding are presented in Table 7.2).

[a]Amounts do not include funding for the DOE Office of Fossil Energy's Carbon Sequestration program, which supports a range of sequestration research and demonstration activities that include geologic sequestration site characterizations.

Coal will continue to provide a major portion of energy requirements in the United States for at least the next several decades, and it is imperative that policy makers are provided with accurate information describing the amount, location, and quality of the coal resources and reserves that will be available to fulfill these energy needs. It is also important that we extract our coal resources efficiently, safely, and in an environmentally responsible manner. A renewed focus on federal support for coal-related research, coordinated across agencies and with the active participation of the states and the industrial sector, is a critical element for each of these requirements.

1

Introduction

Ensuring a stable energy supply for the nation has been a high-priority issue for the U.S. government since the oil embargo of 1973-1974. In the past 30+ years, the nation has experienced energy price controls and decontrols, deregulation of natural gas and electricity, at least three oil price spikes, and one oil price crash. During this time, national energy policy has been created and modified through numerous acts of Congress and executive orders. These policies included the reorganization and consolidation of energy research and development (R&D) activities with the formation of the Energy Research and Development Administration (ERDA), later incorporated into the U.S. Department of Energy (DOE); the dissolution of the U.S. Department of the Interior's Bureau of Mines; and the creation of the U.S. Nuclear Regulatory Commission. Throughout this time, as coal production and use have doubled in response to increased demand for electrical power generation, coal prices have been considerably less volatile than those of other fossil fuels (Box 1.1 and Figure 1.1). Mine safety has been consistently improved; environmental control technologies have reduced emissions of NO_x, SO_2, and particulates from coal-fueled power plants; and the effort to remove abandoned mine land hazards and scars, a vast legacy from earlier coal mining activities, is under way.

Now the nation's policy makers face critical questions. Will coal use continue to increase over the next 25 years, perhaps with increased synfuels production from coal, or will coal use grow at a slower rate—or even decline—if mandatory carbon dioxide emission controls are imposed? Coal technology research, development, demonstration, and deployment policies need to be designed to accommodate a broad range of possible future scenarios. Addressing this significant

13

BOX 1.1
Coal Price Trends

In contrast to price trends for natural gas and oil, which are broadly similar for the period 1949-2005, the price of coal has been much less volatile and has followed a different trajectory (Figure 1.1). A period of decreased coal mine productivity in the mid-1970s, in response to a more constrained regulatory environment, was followed by a long period of decreasing prices resulting from a two- to three-fold increase in productivity. This dramatic productivity increase was largely due to an upsurge in production from large surface mines in the West as well as the consolidation of small mines and the adoption of longwall mining in underground mines in the East. On a constant-dollar basis, the price of coal in 2005 was less than half the price of coal in 1975. On a heating-value basis, oil and gas were several times more expensive than coal in 2005, giving coal a significant price advantage over the competing fossil fuels for use in generating electricity.

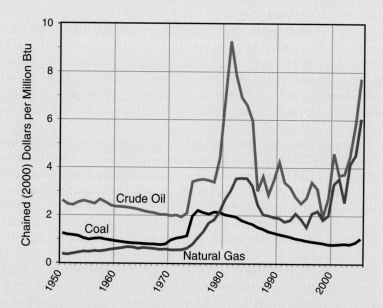

FIGURE 1.1 Fossil fuel production prices for 1949-2005. 'Chained Dollars' are calculated to express real prices relative to a particular reference year (2000 in this case), based on the purchasing power of goods and services in successive pairs of years. SOURCE: EIA (2006a).

challenge was a primary objective of this study—the common thread throughout the committee's deliberations.

COAL IN THE U.S. ENERGY ECONOMY

Different coals have different heating values (energy per unit mass). Therefore, the amount of coal in the overall U.S. energy economy should be considered in terms of both its mass (commonly expressed in tons) and its energy content (commonly expressed in British thermal units, abbreviated as Btu[1]). Annual U.S. coal production has roughly doubled over the past 50 years, and now exceeds 1 billion tons per year (Figure 1.2) (EIA, 2006a). Since the mid-1980s, the proportion of coal in the total U.S. energy mix has remained broadly constant and supplied approximately 23 percent of the 101 quadrillion (10^{15}) Btu of total energy consumed in 2005 (Figures 1.3 and 1.4).

On a tonnage basis, production from large surface mines that are located mostly in the western states (Figures 1.5 and 1.6) has grown rapidly since 1970, while production from underground coal mines, located largely in the interior eastern part of the country, has remained approximately constant (Figure 1.6). Just four states—Wyoming, West Virginia, Kentucky, and Pennsylvania—produce 65 percent of the coal mined in the United States on a tonnage basis. Wyoming supplies almost two and a half times as much coal on a tonnage basis (or about 1.6 times as much on an energy basis) as West Virginia, the next largest coal producer.

Historically, most coal produced in the United States has been consumed in the United States (EIA, 2006c). In 2005, 1.128 billion tons of coal were consumed and 1.133 billons tons were produced. That year, the United States imported 30.5 million tons of coal, mostly from Colombia, and exported 49.9 million tons, with about a third going to Europe and a third going to Canada (EIA, 2006c). Metallurgical coal made up more than half of coal exports (28.7 million tons), primarily to Europe but with lesser amounts going to Canada, Brazil, and Asia (Freme, 2006).

Coal use for electric power generation has risen dramatically in the last half century (Figure 1.7) with most U.S. coal that is produced at present consumed by the electric power sector. That sector alone consumed 1 billion tons of coal in 2005, or 92 percent of all coal produced in the United States that year (EIA, 2006a). Today, coal supplies the energy to produce more than half of the electricity generated in this country, making it a vital part of the U.S. energy economy.

---

[1]Although the standard measure of energy content used by the coal industry in the United States is the Btu, other countries use the International System of Units (metric) measurement system. Unit conversion factors and energy ratings are listed in Appendix G.

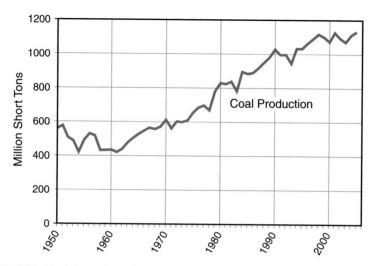

FIGURE 1.2 Total domestic coal production, showing the consistent upward trend since the early 1960s. SOURCE: EIA (2006a).

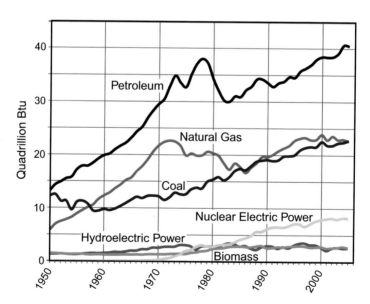

FIGURE 1.3 Total domestic energy consumption by major source, 1949-2005. SOURCE: EIA (2006a).

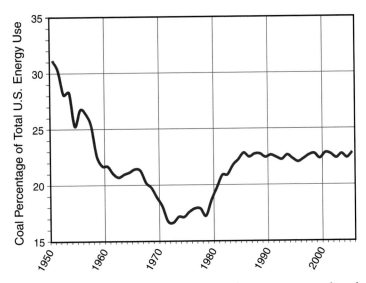

FIGURE 1.4 Coal use as a percentage of total domestic energy consumption, showing that the proportion of coal in the total U.S. energy mix has remained broadly constant since the mid-1980s. SOURCE: Based on data in EIA (2005d).

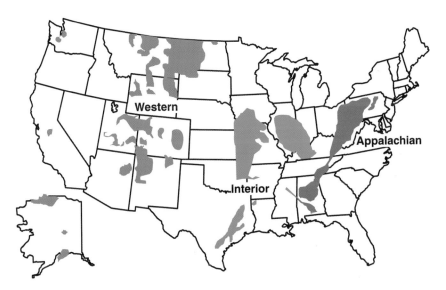

FIGURE 1.5 Major coal-producing regions in the United States. SOURCE: Modified after EIA (2006b).

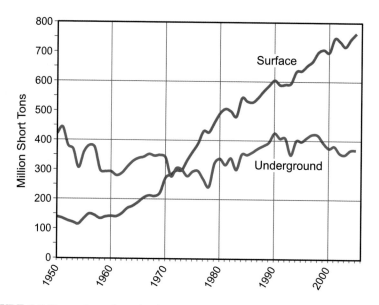

FIGURE 1.6 Domestic coal production since 1950 from surface and underground coal mines, illustrating the dramatic expansion of surface mining (concentrated primarily in the western states). SOURCE: EIA (2006a).

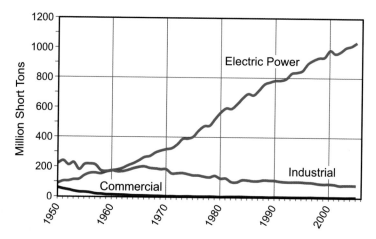

FIGURE 1.7 Historical trends in U.S. coal use by sector, showing the continued and steep rise in coal use for electric power generation as other uses have declined. SOURCE: EIA (2006a).

COMMITTEE CHARGE AND SCOPE OF THIS STUDY

For this report, a broad interpretation of coal R&D has been adopted to include activities that are variously described by different agencies as pure research, applied science, pilot-scale testing, technical support, demonstration projects, and applied engineering projects. Collectively, these research-related activities support the coal component of the federal government's energy portfolio.

The range of agencies and the diversity of federally funded programs raise a number of questions:

- What is the total federal R&D funding across the coal life cycle?
- Have R&D products been successfully integrated into the coal industry?
- Does coal R&D require particular coordination?

In the 2005 Consolidated Appropriations Act (P.L. 108-447), Congress directed the Office of Surface Mining Reclamation and Enforcement (OSM) to contract with the National Research Council to conduct a study on coal research, technology, and resource assessments. The committee's task (Box 1.2) was to broadly examine coal R&D, recognizing that it is an essential component of an appropriate, integrated roadmap for our nation's future energy needs. The analysis would allow policy makers to gauge the success of past research activities, gain a clearer understanding of the research being undertaken throughout the entire coal cycle, and provide updated and expanded information to better prioritize investment and policy needs within the coal sector. By also examining critical gaps in research and technology, and the potential impacts of key policy developments, this study was intended to offer a more complete picture of the role of coal in the U.S. energy mix, and provide the basis for more informed development of a national energy strategy.

To respond to the charge from Congress, the National Research Council established a committee comprising 13 experts with wide-ranging academic, industry, and state government expertise. Committee biographical information is presented in Appendix A. This report is designed for a wide range of audiences. It provides analysis and advice for the U.S. Congress and relevant federal agencies. It is also designed to provide accessible information to other federal agencies, state policy makers, the coal industry, and the general public.

COMMITTEE PROCESS

The committee held seven meetings between January 2006 and February 2007, convening three times in Washington, D.C., and once each in Pittsburgh, Pennsylvania; Spearfish, South Dakota; Boulder, Colorado; and Irvine, California. The committee visited an underground coal mine near Pittsburgh, Pennsylvania, and a surface coal mine in the Powder River Basin near Gillette, Wyoming.

BOX 1.2
Statement of Task

The study will broadly examine coal resource assessments, technologies, and research and development (R&D) activities in the United States in order to formulate an appropriate, integrated roadmap of future needs. The results of the review should help define and construct a national strategy for coal R&D and resource assessments.
The study shall consider the following issues:

1. Summarize recent projections of the coal use as part of the U.S. and global energy portfolios over the next 25 years, including projections that take into account the potential roles of coal in future integrated energy and environmental policies, in order to set the context for development of a more comprehensive, strategic roadmap for coal R&D and resource assessments.
2. Describe the full range of local, regional, national, and global issues and challenges, including environmental issues, that must be taken into account when considering future production and utilization of coal.
3. Review the coal reserve assessments based on recent trends in the coal sector and examine the current and future role of coal imports and exports.
4. Assess the categories of coal R&D currently being carried out in the United States and internationally, and investigate whether and how technology developments in other fields can be applied to the coal sector. Review how technologies are being transferred to coal mine operators and other users, recognizing differences among companies.
5. Determine the priority coal R&D needs, including in the areas of exploration, discovery, reserve assessment (including in terms of commercial feasibility for known reserves), extraction, coal preparation, delivery to market, waste disposal, reclamation, health and safety, community impact, environmental practices, education and training, and productivity.
6. Evaluate the need for a broad-based, coordinated, multi-agency coal research and development program. Review current coal-related research, examine what agencies are conducting it, and determine how much funding is currently being spent throughout the coal life cycle.
7. Examine options for supporting and implementing a broad-based coal R&D program, including approximate costs, and the relative roles and commitments of the public and private sectors now and into the future.

Six of the meetings included information-gathering sessions open to the public. These open sessions included presentations by, and discussions with, representatives from the offices of U.S. Senators Arlen Specter and Robert C. Byrd, and relevant federal government agencies—the U.S. Air Force for the Department of Defense, Office of Advanced Systems and Concepts; the Energy Information Administration (EIA) and National Energy Technology Laboratory (NETL) in the Department of Energy (DOE); the Office of Surface Mining Reclamation and Enforcement (OSM), U.S. Geological Survey (USGS), and Bureau of Land Management (BLM) in the Department of the Interior; the National Institute for Occupational Safety and Health (NIOSH) in the Department of Health and Human Services; the Mine Safety and Health Administration (MSHA) in the Department of Labor; and the U.S. Environmental Protection Agency (EPA). The committee also received briefings by representatives from the International Energy Agency (IEA), industry associations, state organizations, environmental organizations, academic researchers, and labor and industry—these individuals, with their affiliations and presentation titles, are listed in Appendix B. To respond to the statement of task, the committee relied on relevant technical documents, written materials provided to the committee, presentations made to the committee, pertinent National Academies' reports, the committee's observations during mine visits, and the collective expertise of committee members.

Early in the process, the committee queried the all-encompassing nature of the statement of task, which might be interpreted as an invitation to undertake a highly detailed study resulting in a lengthy and comprehensive report covering all aspects of coal production and use. In response, representatives from the offices of Senators Byrd and Specter emphasized to the committee that the advice sought by the congressional mandate was to be broad in scope and insightful, but with limited detail and abundant references to existing more comprehensive studies that address specific topics. Moreover, they indicated that R&D aspects of coal utilization technologies have already been assessed by a range of National Research Council reviews and requested that this study focus primarily on R&D related to all other ("upstream") aspects of the coal fuel cycle. For this reason, the current report presents only a brief overview of coal utilization technologies and related R&D programs. While the committee identifies and highlights a number of critical issues related to coal utilization—in particular, the impact on coal use of government policies regarding climate change and greenhouse gas emissions—it does not evaluate or consider in detail the related R&D programs such as research on carbon capture and sequestration technologies. Rather, in accord with the congressional guidance, coal utilization R&D activities are summarized briefly with references provided to other ongoing programs and assessments.

REPORT ORGANIZATION—THE COAL FUEL CYCLE

The committee used the coal "fuel cycle" as an organizing framework to address the broad scope of the work statement. The fuel cycle is illustrated schematically in Figure 1.8, which depicts the approximate mass flows of coal from reserve assessment, through mining and processing, to end use. Although the energy content per unit of mass varies depending on coal type, the flow of energy embodied in the coal is approximately proportional to the mass flow.

Each stage of the fuel cycle also has associated environmental impacts, in the form of land use requirements and additional flows of wastes or residuals

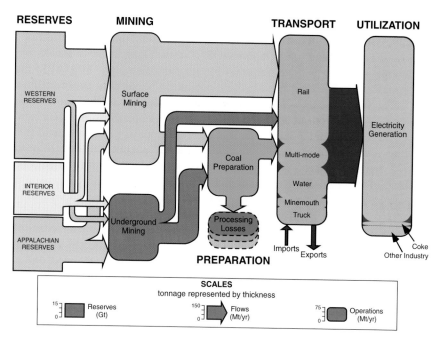

FIGURE 1.8 Schematic showing the coal fuel cycle in United States, illustrating the flow paths and relative quantities of coal as it moves from reserves through the various operations—mining to processing (if applicable) to transport to utilization. The thickness of flow arrows reflects tonnages moved in 2005; similarly, the heights of the reserve and operations boxes reflect tonnage estimates for 2005 (Gt—gigatons; Mt—megatons); note the differing scales for reserves, operations, and flows. The ultimate stage, the distribution of products from the utilization stage, is not depicted. The processing losses box is dashed to reflect the great variability among preparation plants and the difficulty of quantifying losses. SOURCES: concepts and data from Fiscor (2005), NCC (2006), EIA (2006d); Gene Berry (Lawrence Livermore National Laboratory, personal communication, 2006).

emitted to air, land, and water. While the quantities of coal in Figure 1.8 represent the situation in 2005, the future picture could be quite different. For example, some scenarios described in Chapter 2 indicate a potential for substantial growth in the production of coal-derived liquid and gaseous fuel, requiring a transport infrastructure for distributing such products via pipelines.

Chapter 2 first considers the outlook for U.S. and world coal production and use to set the context for this report. The R&D activities associated with each stage in the coal fuel cycle are then discussed more fully in subsequent chapters:

• The first stage of the coal fuel cycle is geological exploration to establish the resource base of coal reserves. Although current estimates of minable coal reserves amount to several hundred years of supply at today's rate of extraction, questions remain regarding the characterization and quantification of coal reserves as well as future rates of coal utilization. Chapter 3 describes coal resource and reserve assessments and addresses issues and concerns associated with these assessments.

- The second stage of the coal fuel cycle is the mining and subsequent processing of coal from underground and surface mines, described in Chapter 4. Coal processing can include a variety of steps—crushing, screening, and wet or dry separations—to reduce the mineral matter (ash) content of coal prior to transport and use. Much western coal is surface mined from very thick (greater than 50 feet) seams of relatively clean coal and shipped after simple crushing and screening, whereas eastern coal—generally mined from thinner seams (less than 10 feet thick)—is characteristically cleaned prior to shipment.

- The third stage of the fuel cycle is the transport of raw or processed coal, described in Chapter 5. Approximately two-thirds of coal production is moved by rail, with trucks, inland waterways, and multimodal transport accounting for the remaining third. One reason for the large market share for rail was the rapid growth in the 1990s of coal production from the Powder River Basin. More than 90 percent of this coal is transported by rail outside the state of origin, with more than 50 percent going to power plants east of the Mississippi River and to Texas. The chapter also includes a brief description of the distribution of coal-derived products to end users, dominated by the electric power transmission system, and discusses the potential future transport of CO_2 captured in coal-fired power plants.

- The fourth stage of the fuel cycle is the conversion of coal to other energy forms, described in Chapter 6. This stage is dominated by the combustion of coal for electric power generation, which accounted for 92 percent of U.S. coal use in 2005. Other major uses of coal are by the industrial sector for the production of coke (used in steel and other metals production processes) and as a boiler fuel to supply process heat and power. As well as a brief analysis of R&D issues associated with coal utilization, Chapter 6 also discusses environmental concerns associated with coal-fired power plants.

Chapter 7 summarizes future projections for coal production and use, notes two important societal issues—community impacts and workforce demographics—that cut across the coal fuel cycle, and presents an outline of current federal support for coal-related R&D. Chapter 7 also summarizes the findings and repeats the recommendations from earlier chapters for additional funding support of upstream R&D activities, and concludes with suggestions for improved coordination of R&D activities among federal agencies, coal-producing states, and the coal industry. As part of its analysis of existing and past coal-related R&D programs and their outputs, the committee used its collective knowledge to provide broad, but necessarily approximate, estimates of the funding levels that will be required to achieve the outcomes described in each recommendation.

2

Projections for U.S. and World Coal Use

This chapter summarizes current projections for U.S. and worldwide coal use over the next 25 years and beyond, and identifies the key factors that influence this outlook. This information provides the context and perspective for later chapters where the components of the coal use fuel cycle are examined in greater detail.

The outlook for future coal production and use presented here is based on recent studies and analyses by government and private organizations in the United States and elsewhere, and results are summarized for a range of scenarios reflecting the key factors that will influence future coal production. Because different organizations employ different methods, assumptions, and scenarios, the results are presented first for each of the major studies or sources of coal production and use projections. Then, the committee's overall findings are presented based on its analysis of the full spectrum of studies reviewed.

COAL PRODUCTION SCENARIOS FOR THE UNITED STATES

This section summarizes estimates of future U.S. coal production and use for a range of scenarios developed by the Department of Energy's Energy Information Administration (DOE-EIA) and the Pacific Northwest National Laboratory (PNNL). These scenarios reflect a range of assumptions about technical, economic, and policy variables that will influence future coal production and use; they are intended to be illustrative of recently published work by a variety of public and private organizations involved in energy and environmental modeling.

U.S. Energy Information Administration Projections

A principal source for projections related to energy use in the United States is the *Annual Energy Outlook* (AEO), updated each year by the EIA. The AEO is based on the National Energy Modeling System (NEMS) developed by the EIA and used to project energy use over the next 25 years for a range of scenarios. The "reference case" scenario is arguably the most widely cited of EIA cases. It reflects EIA's best estimate of trends for a "business-as-usual" case that assumes continuation of all current laws, regulations, and policies. Other scenarios published by EIA use different assumptions about factors such as economic growth rates, fuel cost or price trends, and rates of technological change in different sectors of the economy (Table 2.1).

TABLE 2.1 Summary of Cases Used in EIA 2006 Projections of U.S. Coal Use Assuming No Change in Current Policies

Scenario	Description
Reference case	Baseline economic growth (3%/yr), increased world oil price, and assumptions about adoption of renewable, nuclear, and other energy technologies. Gradual decline of minemouth coal price
Low coal cost	Productivity for coal mining and coal transportation assumed to increase more rapidly than in the reference case. Coal mining wages, mine equipment, and coal transportation equipment costs assumed to be lower than in the reference case
High coal cost	Productivity for coal mining and coal transportation assumed to increase more slowly than in the reference case. Coal mining wages, mine equipment and coal transportation equipment costs assumed to be higher than in the reference case
Low economic growth	Gross domestic product grows at an average annual rate of 2.4% for 2004 through 2030
High economic growth	Gross domestic product grows at an average annual rate of 3.5% for 2004 through 2030
Low O&G price	Prices for worldwide crude oil and natural gas (O&G) resources are lower than in the reference case. World oil prices are $28 per barrel in 2030, compared to $50 per barrel in the reference case, and lower-48 wellhead natural gas prices are $4.96 per thousand cubic feet in 2030, compared to $5.92 in the reference case.
High O&G price	Prices for worldwide crude oil and natural gas resources are higher than in the reference case. World oil prices are about $90 per barrel in 2030 and lower-48 wellhead natural gas prices are $7.72 per thousand cubic feet in 2030, compared to $5.92 in the reference case.
Slow O&G technology	Cost, finding rate, and success rate parameters adjusted for 50% slower improvement than in the reference case
Rapid O&G technology	Cost, finding rate, and success rate parameters adjusted for 50% more rapid improvement than in the reference case

SOURCE: EIA (2006d).

The EIA is precluded from analyzing alternative policy scenarios as part of the AEO. For example, the AEO does not include any cases in which U.S. greenhouse gas emissions are constrained over the next 25 years, since there is currently no policy that restricts such emissions. However, EIA does publish the results of policy analysis studies performed at the request of members of Congress, and these studies provide an important complement to the AEO because they explore a wider range of factors relevant to energy use projections. Table 2.2 shows additional EIA cases developed recently for a congressionally requested

TABLE 2.2 GHG Policy Cases Modeled by the EIA for Congressionally Requested Studies

Case Name	GHG Intensity Reduction Goal (% per year)[a]		Safety-Valve Price (2004 dollars per tonne CO_2 equivalent)[b]		Description[b]
	2010-2019	2020-2030	2010	**2030**	
Cap-Trade 1	2.4	2.8	$6.16	$9.86	
Cap-Trade 2	2.6	3.0	$8.83	$14.13	GHG cap-and-trade
Cap-Trade 3	2.8	3.5	$22.09	$35.34	system with safety valve
Cap-Trade 4	3.0	4.0	$30.92	$49.47	
Cap-Trade 3 Low Other	2.8	3.5	$22.09	$35.43	Cap-Trade 3 with 50% reduction in "other than energy-related CO_2 GHG abatement"
Cap-Trade 3 Low Safety	2.8	3.5	$8.83	$14.13	Cap-Trade 3 with lower assumed safety-valve price
Cap-Trade 3 High Tech	2.8	3.5	$22.09	$35.34	Cap-Trade 3 with more optimistic technology assumptions

NOTE: These scenarios are illustrative of a range of policy proposals that would limit emissions of CO_2 from coal combustion.

[a]GHG intensity refers to annual GHG emissions per dollar of gross domestic product for a given year.

[b]A cap-and-trade program places an overall limit on total GHG emissions from all emission sources in a given year. The annual cap is determined by the required GHG intensity reduction. Each source is required to hold one emissions "allowance" for each ton emitted, with the total number of annual allowances set by the government to be equal to the total tons in the cap. Allowances may be freely traded, offering sources the option of complying either by reducing emissions, by buying more allowances in the market, or by a combination of both strategies. The "safety valve" allows total emissions to exceed the cap if the market price for allowances exceeds the specified safety-valve price. In effect, the safety-valve price is the maximum price for allowances in the market. All permit safety-valve prices shown in Table 2.2 are in 2004 dollars. The range requested for this study was $10 to $35 in 2010 dollars (corresponding to $8.83 to $30.92 in 2004 dollars shown in the table). The safety valves are assumed to increase by 5 percent annually in nominal dollars from 2010 through 2030.

SOURCE: EIA (2006e).

study of alternative cap-and-trade policies that would restrict U.S. greenhouse gas (GHG) emissions over the next several decades. These cases explore different levels of reduction in GHG intensity (defined as GHG emissions per unit of gross domestic product), beginning with the level proposed by the National Commission on Energy Policy (NCEP, 2004). These scenarios are illustrative of a variety of congressional proposals that would limit carbon dioxide emissions from fossil fuel combustion. Such scenarios are especially relevant to the present study since they explore the impact of policy measures that directly affect future U.S. coal production and use. Figure 2.1 shows the trends in GHG emissions associated with the scenarios in Table 2.2.

Figure 2.2 summarizes the range of total coal use projections in British thermal units (Btu) for the years 2020 and 2030 as reported by EIA, and Figure 2.3 summarizes the corresponding range of regional coal production figures (in units of tons rather than energy) projected by EIA for each of the scenarios shown in Tables 2.1 and 2.2. The results of these cases show a very wide range of future U.S. coal production estimates. Relative to the reference case scenario, which projects an approximately 50 percent increase in coal energy use by 2030 (relative to 2004), sustained high oil and gas prices yield an even greater increase of about 70 percent above present levels. The latter scenarios, however, assume no future constraints on GHG emissions. In contrast, scenarios that do limit future GHG emissions show dramatically different results. In these scenarios, coal use is curtailed significantly and falls below 2004 levels in the most restrictive cases. Coal production in the western states is impacted more severely than eastern coal

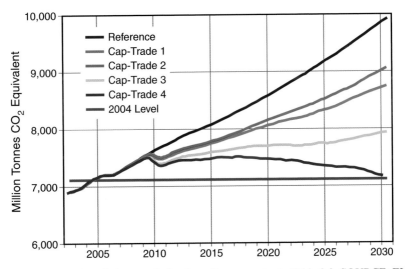

FIGURE 2.1 GHG emission trends for the policy scenarios in Table 2.2. SOURCE: EIA (2006e).

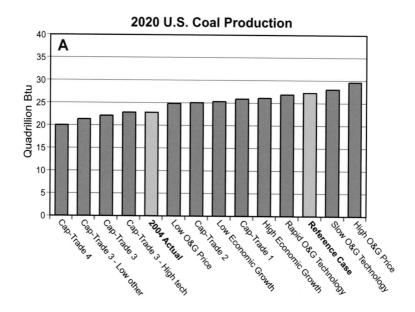

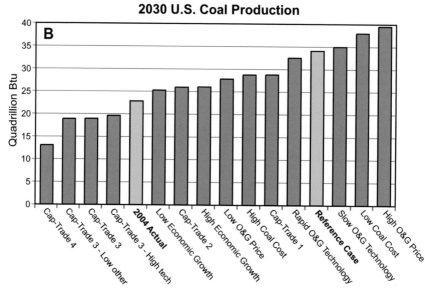

FIGURE 2.2 Projections by the U.S. Energy Information Administration for U.S. coal use in (A) 2020 and (B) 2030 for the range of scenarios listed in Tables 2.1 and 2.2. SOURCES: EIA (2006d, 2006e).

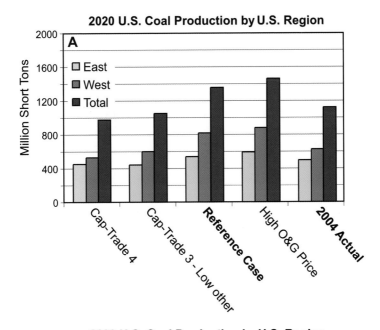

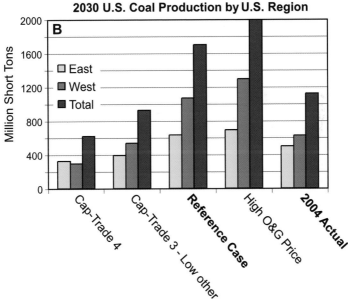

FIGURE 2.3 Projections by the U.S. Energy Information Administration for U.S. coal use, on a tonnage basis, east and west of the Mississippi River in 2020 and 2030 for the range of scenarios listed in Tables 2.1 and 2.2. SOURCES: EIA (2006d, 2006e).

production, because most of the growth in the reference case is projected to occur in the West in the absence of a GHG emission constraint.

The impacts of these EIA scenarios on projected changes in coal imports and exports are summarized in Table 2.3. U.S. coal imports have been increasing at a relatively rapid rate, from less than 10 million tons in 2000 to more than 30 million tons in 2005. This trend has been driven mainly by the low sulfur content and lower delivered cost to the eastern U.S. markets of coals from Colombia, Venezuela, and Indonesia. Coal imports are expected to increase further over current levels and to exceed declining levels of exports by 2020 (Table 2.3). While the magnitude of imports and exports remains small relative to total coal use, the relative changes in Table 2.3 project that imports will increase by 50 to 240 percent while exports decline by 47 to 68 percent. In its reference case scenario, EIA projects that in 2030 the United States will import 91 million tons (2.37 quadrillion Btu) of coal, approximately three times as much as in 2005 (EIA, 2006d). In contrast, current U.S. coal exports are about 50 million tons per year, a little less than half of the record export tonnages in the 1980s. Exports are expected to decrease in the future, primarily due to the anticipated availability of low-cost coal supplies from South America, Asia, and Australia. The EIA predicts that the U.S. share of total world trade will fall from 6 percent in 2003 to 3 percent in 2025.

The range of energy sources for electric power generation is shown in Table 2.4; the three cases shown span the range of coal use projections in Tables 2.1 and 2.2. Significant changes in the amount of coal used for power generation also are seen in Table 2.4. In the AEO high oil and gas price scenario (which gives the largest increase in coal production), the fraction of electricity generated from coal climbs from its current (2004) share of 50 percent to 55 percent in 2020 and 64 percent in 2030. These values are slightly greater than the reference case scenario in the absence of a carbon constraint. However, under the Cap-Trade 4 scenario, coal's share of electricity production declines to 37 percent in 2020 and

TABLE 2.3 Actual and Projected Coal Imports and Exports for Selected EIA Scenarios

Scenario	Actual 2004		2020		2030	
	Imports	Exports	Imports	Exports	Imports	Exports
Cap-Trade[a]	0.79	1.25	1.25	0.72	1.19	0.66
AEO Reference Case[a]	0.79	1.25	1.31	0.46	2.37	0.40
AEO High O/G[b]	0.79	1.25	1.61	0.46	2.69	0.40

NOTE: All values in quads (quadrillion Btu).

[a]Imports include coal and coke (net).
[b]Imports include coal, coke, and electricity (net).

SOURCES: EIA (2006d, 2006e).

TABLE 2.4 Actual and Projected Relative Contributions of Different Energy Sources for Electricity Production for Selected EIA Scenarios

Scenario	Actual 2004		2020		2030	
	Quad Btu	% of Total	Quad Btu	% of Total	Quad Btu	% of Total
Cap-Trade 4						
Coal	20.26	52.4	18.49	40.1	11.63	22.6
Nuclear	8.23	21.3	9.86	21.4	18.39	35.8
Gas and Oil	6.57	17.0	6.70	14.5	6.21	12.1
Renewable & Other	3.60	9.3	11.03	23.9	15.21	29.6
Total	38.67		46.09		51.44	
AEO Reference Case						
Coal	20.26	52.4	25.02	51.9	30.74	57.2
Nuclear	8.23	21.3	9.09	18.8	9.09	16.9
Gas and Oil	6.57	17.0	8.62	17.9	7.61	14.2
Renewable & Other	3.60	9.3	5.52	11.4	6.27	11.7
Total	38.67		48.24		53.71	
AEO High O/G Prices						
Coal	20.26	52.4	27.30	56.1	32.57	61.3
Nuclear	8.23	21.3	9.09	18.7	9.09	17.1
Gas and Oil	6.57	17.0	6.27	12.9	5.11	9.6
Renewable & Other	3.60	9.3	5.98	12.3	6.37	12.0
Total	38.67		48.64		53.14	

SOURCES: EIA (2006d, 2006e).

22 percent in 2030, according to EIA models. The shares of nuclear and renewable energy increase significantly in that scenario.

While EIA scenarios are widely cited and provide detailed information that is publicly available, other organizations also publish energy forecasts or scenarios.

Pacific Northwest National Laboratory Projections

Given the importance of carbon constraints and fuel prices to projections for future coal use revealed by the EIA scenarios, the committee presents recent findings from Pacific Northwest National Laboratory that project future U.S. coal use for a longer period of time under different policy scenarios. PNNL, one of the 17 national government research laboratories supported by DOE, has the mandate to conduct research and develop technology to support DOE's Office of Science and other DOE offices. Large-scale energy models developed at PNNL have been used extensively to analyze alternative energy futures and policy scenarios for GHG reductions, both globally and domestically. Recent PNNL studies examined the effects of carbon constraints and fuel prices on future U.S. coal use from

2005 to 2045 (Wise et al., 2007). The two carbon price (CP) scenarios modeled by PNNL assumed future market prices for CO_2 allowances growing at different rates. In one case (called CP1), allowance prices per tonne of CO_2 increased from an initial $12 in 2015 to $20 by 2035 and $25 in 2045. The second case (CP2) started at the same $12 per tonne in 2015, but increased more sharply to $32 by 2035 and $52 in 2045. Each CP case was combined with two fuel price (FP) cases based on EIA's Annual Energy Outlook. The "base case" fuel prices (FP1) were the same as EIA's Reference Case values, while the second case (FP2) represented EIA's "constrained supply" case in which natural gas prices for power generation rise to $7 to $9 per million Btu (MBtu) by 2030 (compared with $5 to $7/MBtu for the base case). Figure 2.4 shows the resulting CO_2 emissions from U.S. power plants for the four scenarios, and Figure 2.5 shows the impact on utility coal use, expressed in terms of the installed capacity of coal-fired power plants. Also shown is a "business-as-usual" reference case, which uses the EIA's base case energy prices but does not impose any CO_2 emissions control policy (CP0FP1).

Figures 2.4 and 2.5 show that the lower CO_2 allowance prices (up to about $25 per tonne of CO_2) result in reduced CO_2 emissions, as well as decreased use of coal, relative to the base case, as in the EIA cap-and-trade scenarios shown earlier. However, for higher natural gas prices (FP2 cases), coal use actually exceeds the reference case value, even with higher carbon prices. In these scenarios, carbon capture and sequestration (CCS[1]) plays an increasingly important role in reducing CO_2 emissions and enabling coal to remain economically viable. The combination of high carbon prices and high natural gas prices (scenario CP2FP2) brings about the largest long-term reduction in CO_2 emissions as well as the greatest increase in coal use—exceeding even the business-as-usual (CP0FP1) reference case scenario projections. In large part, this is because the PNNL scenarios have much smaller increases in the use of nuclear and renewable energy for power generation compared to the earlier EIA Cap-Trade 4 scenario. Most of the fuel substitutions in the PNNL cases occur between coal (with and without CCS) and natural gas (Wise et al., 2007).

INTERNATIONAL COAL PRODUCTION PROJECTIONS

This section summarizes results of scenarios developed by a number of different organizations to estimate future international coal production and use. Again, results are intended to be illustrative of the range of technical, economic, and policy variables that will influence future coal production and use.

---

[1]Also often expressed as Carbon Capture and Storage.

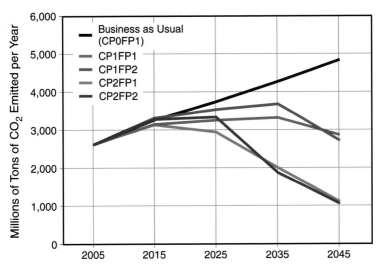

FIGURE 2.4 Effect of different carbon and fuel price scenarios on annual CO_2 emissions from electricity generation for PNNL scenarios. SOURCE: Wise et al. (2007).

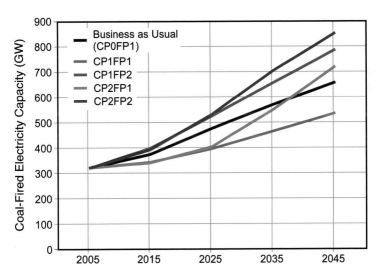

FIGURE 2.5 Effect of different carbon and fuel price scenarios on coal-fired capacity for electricity generation in gigawatts (GW) for PNNL scenarios. SOURCE: Marshall Wise and James Dooley, Joint Global Change Research Institute, Pacific Northwest National Laboratory, personal communication, 2007.

U.S. Energy Information Administration Projections

Along with domestic projections, the EIA also publishes scenarios for international energy use in its annual *International Energy Outlook*. Figure 2.6 summarizes the most recent EIA projections of world coal use in 2010, 2020, and 2025, for three groups of countries—mature market economies (including most Organisation for Economic Co-operation and Development [OECD] countries), emerging economies (such as China and India), and transitional economies (including the former Soviet Union, non-OECD Europe, and Eurasia). Again, these scenarios reflect variations in different growth rate parameters, but do not include policy scenarios such as future GHG constraints. In the absence of such policy constraints, world coal use is projected to grow dramatically in the emerging economies, primarily China and India. Much smaller tonnage growth is projected in the rest of the world, although relative growth rates are projected to be high in several other countries. By 2025, worldwide coal use increases by approximately 60 percent over 2002 levels in the reference case and by nearly 80 percent in the high economic growth scenario.

World Energy Council Projections

The World Energy Council (WEC) is an independent organization that draws on national studies and data from member countries to project worldwide energy consumption. Recently, it developed six scenarios for future global primary energy supply and the associated carbon dioxide emissions (expressed as emissions of carbon) to 2050 (WEC, 2006): Case A1, high growth with emphasis on increased use of oil and gas; Case A2, high growth and coal intensive; Case A3, high growth with emphasis on natural gas, new renewables, and nuclear energy; Case B, a middle-course reference case most often cited for comparison purposes; Case C1, an "ecologically driven" climate policy scenario involving carbon constraints together with a phase-out of nuclear energy; and Case C2, which is similar to C1 but with nuclear power playing an expanded role.

The major factors considered in the WEC projections are world population, world economic growth, and world primary energy intensity. The product of these three factors results in a primary energy demand that is from 1.7 to 2.8 times greater in 2050 than the 1990 world energy demand (Table 2.5). An absolute reduction in coal consumption is projected for the two scenarios (C1, C2) in which a carbon dioxide emission constraint is included. For these two scenarios, worldwide carbon emissions in 2050 fall to below 1990 levels. A third scenario, which emphasizes natural gas, renewables, and nuclear energy (A3), results in no gain or loss in the amount of coal utilized in 2050. The remaining cases project world coal use to roughly double (A1, B) or nearly quadruple (A2) in the absence of a carbon constraint.

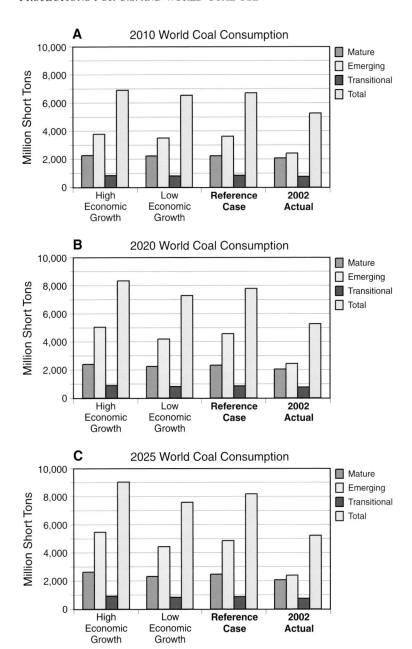

FIGURE 2.6 Projections by the U.S. Energy Information Administration for world coal use in 2010, 2020, and 2025 for the reference case and for high- and low-economic-growth scenarios. SOURCES: EIA (2006d, 2006e).

TABLE 2.5 Projections of the Components of Global Primary Energy Supply[a] and Carbon Emissions[b] in 2050 for Six Scenarios, Compared to 1990 Values

	1990 Value	2050 Scenarios					
		A1	A2	A3	B	C1	C2
Coal	2.2	3.8	7.8	2.2	4.1	1.5	1.5
Oil	3.1	7.9	4.8	4.3	4.0	2.7	2.6
Gas	1.7	4.7	5.5	7.9	4.5	3.9	3.3
Nuclear	0.5	2.9	1.1	2.8	2.7	0.5	1.8
Hydro	0.4	1.0	1.1	1.1	0.9	1.0	1.0
New renewables	0.2	3.7	3.8	5.7	2.8	3.8	3.2
Traditional biomass	0.9	0.8	0.7	0.8	0.8	0.8	0.8
Total (Gtoe)	**9.0**	**24.8**	**24.8**	**24.8**	**19.8**	**14.2**	**14.2**
Carbon emissions (GtC)	**6.0**	**11.7**	**15.1**	**9.2**	**10.0**	**5.4**	**5.0**

[a]Gigatons of oil equivalent (Gtoe).
[b]Gigatons of carbon (GtC).

SOURCE: Used by permission of World Energy Council, London, http://www.worldenergy.org.

ExxonMobil Projections

ExxonMobil Corporation prepares an annual energy outlook that currently presents projections to 2030 (ExxonMobil, 2005). The primary drivers for its energy projections are population growth and gross domestic product (GDP). ExxonMobil estimates a rapid growth in GDP in the developing countries, especially China and India. Secondary factors are efficiency improvements (which reduce energy intensity), changing trends in future consumption patterns, and competition between fuels and available supply. This outlook does not include a carbon constrained case. The growth rate for total energy from 2003 to 2030 is projected to be 1.6 percent, and the growth rate for coal for that period is slightly higher at 1.8 percent per annum. This gain in coal production would result in a 62 percent increase in world coal utilization in the 27 years from 2003 to 2030.

The increase in world coal consumption is also projected on a regional basis, with coal demand in North America and Europe increasing at a modest annual rate of 0.4 and 0.1 percent, respectively, while the Asia Pacific region increases at a much greater rate of 3.1 percent per annum driven by economic growth and large indigenous coal resources.

International Energy Agency Projections

The International Energy Agency (IEA) regularly monitors global energy developments and periodically publishes a *World Energy Outlook* (WEO) (e.g., IEA, 2006a) as well as other special studies. Its most recent study uses the IEA Energy Technology Perspectives model to project world energy use to 2050 for six cases called the Accelerated Technology (ACT) and TECH Plus scenarios that are intended to reduce the growth in global CO_2 emissions relative to the IEA Baseline Scenario (IEA, 2006b). Figure 2.7 describes the nature of the six energy technology scenarios. "ACT Map" refers to an accelerated technology development scenario that is relatively optimistic across all technology areas and results in stabilization of future CO_2 emissions; the remaining scenarios are compared (or "mapped") to the ACT Map scenario. Figure 2.8 shows the resulting effect on global CO_2 emissions, and Figure 2.9 shows the impacts on global coal use relative to the WEO reference case scenario.

	Renewable	Nuclear	CCS	H₂ Fuels	Advanced Biofuels	End-Use Efficiency
Map						
Low Renewables	Pessimistic					
Low Nuclear		Pessimistic				
No CCS			No CCS			
Low Efficiency						Pessimistic
TECH Plus	Optimistic	Optimistic		Optimistic	Optimistic	

(Header spanning all technology columns: TECHNOLOGIES)

FIGURE 2.7 Overview of scenario assumptions for the International Energy Agency ACT and TECH Plus scenarios compared to the ACT Map scenario. SOURCE: *Energy Technology Perspectives* © IECD.UEAM 2996, Table 2.1, p. 43.

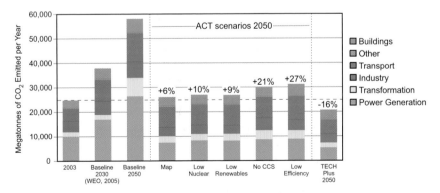

FIGURE 2.8 International Energy Agency projections of global CO_2 emissions for the Baseline, ACT, and TECH Plus scenarios. SOURCE: *Energy Technology Perspectives* © IECD.UEAM 2996, Figure 2.1, p. 46 (as modified).

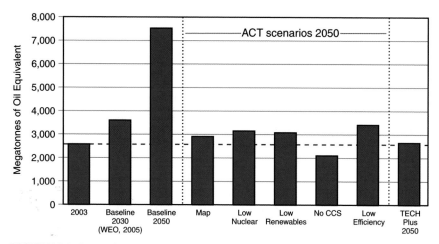

FIGURE 2.9 International Energy Agency projections of global coal use for the Baseline, ACT, and TECH Plus scenarios. SOURCE: *Energy Technology Perspectives* © IECD. UEAM 2996, Figure 2.12, p. 66 (as modified).

In the IEA Baseline Scenario, world coal use in 2050 is nearly three times greater than in 2003, and its share of world energy demand grows from 24 percent in 2003 to 34 percent in 2050. With the accelerated technology scenarios, the increase in coal use is much smaller, exceeding 2003 levels by no more than about 25 percent by 2050. In the absence of CCS technology, 2050 coal use falls below 2003 levels. In all but one of the six technology scenarios in Figure 2.8, global CO_2 emissions still exceed 2003 levels but are sharply reduced relative to the Baseline Scenario.

European Commission Projections

In 2003, the European Commission (EC) published *World Energy, Technology, and Climate Policy Outlook* (WETO), an extensive analysis that includes projections to 2030 for the use of all forms of energy (EC, 2003). The EC developed two scenarios—a business-as-usual case (the reference scenario) and a carbon abatement scenario.

For the reference scenario, worldwide energy demand in 2030 is projected to be 17.1 Gtoe (gigatons of oil equivalent), based on a 1.8 percent annual rate of increase. For the carbon abatement scenario, the energy demand is projected to be 15.2 Gtoe, based on an annual 1.3 percent rate of increase. The worldwide demand for coal in 2030 is projected to be 4.7 Gtoe in the reference scenario and 2.7 Gtoe in the carbon abatement case. These projections indicate that world coal consumption would nearly double from 2000 to 2030 in the reference case,

TABLE 2.6 Comparison of Projections for World Coal Demand by Different Organizations

Scenario	Growth Rate (% per year) for the Indicated Period			World Coal Demand (Mtoe) in Indicated Year			
	2000-2010	2010-2020	2020-2030	2000	2010	2020	2030
WETO	2.07	2.42	2.48	2,389	2,931	3,723	4,757
DOE	1.88	1.50			2,878	3,340	
IEA	1.74	1.74			2,837	3,370	
WEC A2	2.13	2.31	2.22		2,949	3,707	4,616

NOTE: None of these scenarios include a constraint on future CO_2 emissions.

SOURCE: European Commission, World energy, technology and climate policy outlook-2030; Office for Official Publications of the European Communities, Luxembourg, 2003 (EC, 2003).

but would increase by only 13 percent in the carbon abatement case. The WETO projection for primary coal production in North America in 2030 is 1,011 Mtoe (million tons of oil equivalent), representing a growth rate of 1.7 percent per annum from 2010 to 2030 in the absence of a carbon constraint.

The European Commission report compares the WETO projections to those made by the EIA, IEA, and WEC (using the WEC high-growth, coal-intensive scenario A2). The projections from the four agencies do not vary significantly (Table 2.6), and the two that project to 2030 both indicate a doubling of worldwide coal utilization by that time. There would be a much greater difference in projected coal consumption estimates if scenarios with carbon abatement had been included.

Projections by the Intergovernmental Panel on Climate Change

The Intergovernmental Panel on Climate Change (IPCC) has examined a broad range of world energy scenarios (IPCC, 2001) and used selected scenarios in a recent evaluation of CCS and renewable energy sources as a potential climate change mitigation measure for fossil fuel power plants and other major CO_2 sources (IPCC, 2005). IPCC comparison of results from two large-scale models used to project future world energy trends shows that both models project future coal use to remain relatively constant, with increasing use of CCS over time, for a policy scenario aimed at stabilizing atmospheric CO_2 concentrations at 550 parts per million by volume (approximately twice the pre-industrial level) by the end of this century (IPCC, 2005).

HOW WELL DO MODELS PREDICT REALITY?

Comparisons of actual coal production in the United States with projections produced by the U.S. Energy Information Administration illustrate the uncertainties inherent in such forecasts. EIA reference case forecasts are based on conditions prevailing at the time they were made and do not take into account alternative policy scenarios such as those discussed above.

EIA coal production projections made during the period of rapid growth between 1982 and 1989 significantly overestimated actual production, as well as rates of production increases, over a 10-year period (Figure 2.10A). Coal production projections were more realistic during the 1990 to 1993 period of recession and more pessimistic in 1994 and 1995 after three years of rapid decrease in production (Figure 2-10B). Projections made between 1996 and 2004 overestimated production during and following a period of sustained economic growth (1994 to 1998) (Figure 2-10C). Data were not available to compare actual versus projected coal production over longer periods (e.g., 25 years).

These historical EIA reference case projections indicate that there is a tendency to overestimate future production when production is rapidly increasing and to underestimate future production when production is decreasing. When projections are made 10 years ahead, these estimation errors are of the order of 50 to 100 million tons per year of coal production, or approximately 5 to 10 percent of total U.S. production. These errors are likely to increase when longer periods and other scenarios are considered. Thus, while the trends predicted by the future scenarios described earlier are indicative of how coal production may be influenced by various factors, actual values could be significantly higher or lower than projected.

FINDINGS—PROJECTIONS FOR FUTURE COAL PRODUCTION AND USE

While many factors will affect the future use of coal in the United States and globally over the next 25 years or more, recent analyses of coal production and use over the next few decades indicate the following key conclusions:

• Projections show that future coal use depends primarily on the timing and magnitude of potential regulatory limits on CO_2 emissions, on the future demand for electricity, on the prices and availability of alternative energy sources for electric power generation, and on the availability of carbon capture and sequestration technology.

• Over the next 10 to 15 years (until about 2020), coal production and use in the United States are projected to range from about 25 percent above to about 15 percent below 2004 levels, depending on economic conditions and environmental policies. By 2030, the range of projected coal use in the United States broadens considerably, from about 70 percent above to 50 percent below current levels.

• At present, coal imports and exports represent small fractions of total U.S. coal production and use. Projections indicate that imports and exports are expected to remain relatively small.

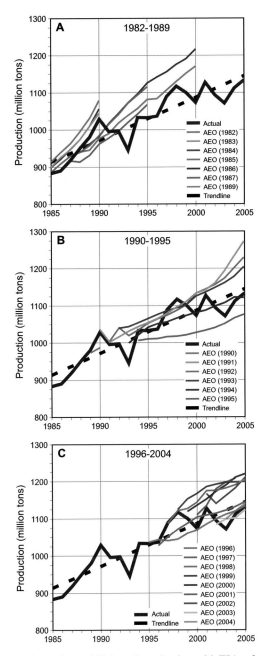

FIGURE 2.10 Comparison of actual U.S. coal production with EIA reference case projections: (A) between 1982 and 1989, (B) between 1990 and 1995, and (C) between 1996 and 2004. SOURCE: EIA (2007a).

• Globally, the largest tonnage increases in coal use are expected in the emerging economies of China and India. Much smaller tonnage growth is projected in the rest of the world, although relative grow rates are projected to be high in several other countries. Again, however, there is great uncertainty in global coal use projections, especially beyond about 2020.

3

Coal Resource, Reserve, and Quality Assessments

Federal policy makers require sound coal reserve data in order to formulate coherent national energy policies. Accurate and complete estimates of national reserves are needed to determine how long coal can continue to supply national electrical power needs, and to determine whether coal has the potential to replace other energy sources, such as petroleum, that may be less reliable or less secure. The coal production and utilization industries—as well as the transportation industry, equipment manufacturers and suppliers, engineering and environmental consultants, federal and state policy makers, financial institutions, and electric transmission grid planners and operators—all require accurate coal reserve estimates for planning. The location, quantity, and quality of coal reserves are critical inputs for determining where end-user industries should be located and for understanding the infrastructure (e.g., trains, barges, haul roads, silos, preparation facilities, power plants, pipelines, electrical transmission lines) that will be needed to support coal production and use. An accurate, comprehensive assessment of the nation's coal resources is essential for informed decisions that will allow the use of this resource so that negative environmental and human health impacts are minimized and there can be an orderly transition from one region to another as the reserves being mined are exhausted. Any substantial increase—perhaps even doubling of current coal production and utilization, as implied by some of the scenarios presented in Chapter 2—will spawn technological, economic, social, environmental, and health issues that will be better anticipated and more efficiently addressed if the location, quantity, and quality of the coal that will be mined over the next several decades are known.

The United States is endowed with a vast amount of coal. The U.S. Geological Survey (USGS) estimated that there are nearly 4 trillion tons of total coal resources in the United States. (Averitt, 1975). However, this estimate has little practical significance because most of this coal cannot be mined economically using current mining practices. A more meaningful figure is the ~267 billion tons of Estimated Recoverable Reserves (ERR) (EIA, 2006a) that is the basis for the commonly reported estimate that the United States has at least 250 years of minable coal.[1] This chapter addresses two major questions to place existing estimates of the amount of usable coal into a broad perspective:

1. Are estimates of available coal reliable, and are they good enough to allow federal policy makers to formulate coherent national energy policies?
2. Can coal reserves in the United States produce the total 1.7 billion tons per year of coal required in 2030 if the Energy Information Administration (EIA) reference case described in Chapter 2 becomes a reality?

The answer to the second question, whether the United States has enough minable coal to meet the projected demands in the EIA reference case, is definitely yes. Coal mining companies report at least 19 billion tons of Recoverable Reserves at Active Mines (EIA, 2006a), and the coal industry reports about 60 billion tons of reserves held by private companies (NMA, 2006a). If recoverable reserves on private, federal, and state lands are added, there is no question that sufficient minable coal is available to meet the nation's coal needs through 2030. Looking further into the future, there is probably sufficient coal to meet the nation's needs for more than 100 years at current production levels. However, it is not possible to confirm that there is a sufficient supply of coal for the next 250 years, as is often asserted. A combination of increased rates of production with more detailed reserve analyses that take into account location, quality, recoverability, and transportation issues may substantially reduce the estimated number of years supply. This increasing uncertainty associated with the longer-term projections arises because significant information is incomplete or unreliable. The data that are publicly available for such projections are outdated, fragmentary, or inaccurate—these deficiencies are elaborated below. Because there are no statistical measures to reflect the uncertainty of the nation's estimated recoverable reserves, future policy will continue to be developed in the absence of accurate estimates until more detailed reserve analyses—which take into account the full suite of geographical, geological, economic, legal, and environmental characteristics—are completed.

RESOURCE AND RESERVE DEFINITIONS

The terms *coal resources* and *coal reserves* are commonly misused and mistakenly interchanged. *Coal resource* is a more general term that describes

---

[1]This statistic is derived by dividing the Energy Information Administration figure of 267 billion tons of ERR by the current annual coal usage of approximately 1.1 billion tons.

naturally occurring deposits in such forms and amounts that economic extraction is currently or potentially feasible (Wood et al., 1983). *Coal reserve* is a more restrictive term describing the part of the coal resource that can be mined economically, at the present time, given existing environmental, legal, and technological constraints (Wood et al., 1983). Coal reserve estimates are often considered the more important parameter because they quantify the amount of recoverable coal. However, coal resource estimates are also important because they are the basis for reserve estimates, and in areas where the data required for defining reserves are missing or inadequate, they provide an indication of the amount of coal in the ground.

The coal resource and reserve classification system currently in use in the United States (Figure 3.1) has undergone more than a century of development. The current system was adopted in 1976 by the U.S. Geological Survey and the U.S. Bureau of Mines (USDOI, 1976) and modified in USGS Circular 891 (Wood et al., 1983). Circular 891 established a uniform foundation for coal resource and reserve assessments by providing standard definitions, criteria, guidelines, and methods. Circular 891 defined coal resource and reserve classes according to their degree of geological reliability (horizontal axis) and economic feasibility (vertical axis) (Figure 3.1), with reliability categories based on the distance from data

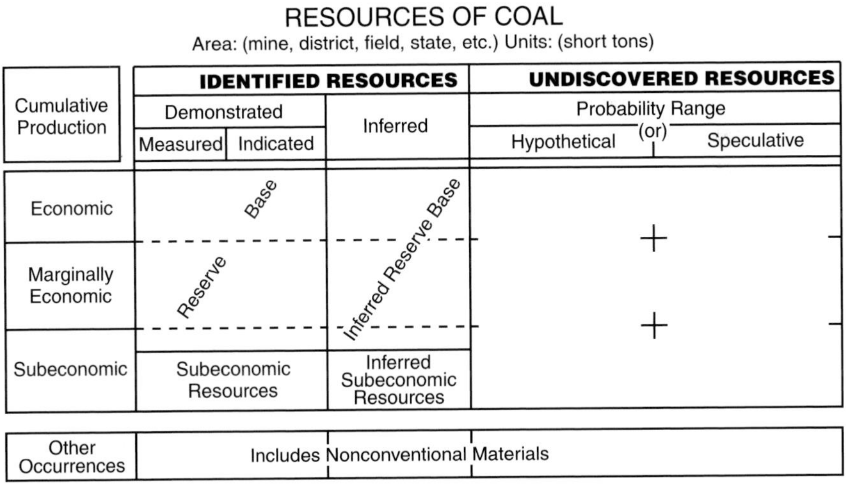

RESOURCES OF COAL
Area: (mine, district, field, state, etc.) Units: (short tons)

Cumulative Production	IDENTIFIED RESOURCES			UNDISCOVERED RESOURCES	
	Demonstrated		Inferred	Probability Range	
	Measured	Indicated		Hypothetical (or)	Speculative
Economic	*Base*		*Reserve Base*	+	
Marginally Economic	*Reserve*		*Inferred*		
Subeconomic	Subeconomic Resources		Inferred Subeconomic Resources	+	
Other Occurrences	Includes Nonconventional Materials				

FIGURE 3.1 Definition of coal resource and reserve classes based on the geological reliability (horizontal axis) and economic viability (vertical axis) of resource estimates. This diagram, often referred to as the McKelvey diagram after a former director of the USGS, represented the state of the art for resource depiction at the time of its publication. SOURCE: Wood et al. (1983).

points (¼ mile for "measured," ¾ mile for "indicated," 3 miles for "inferred," and >3 miles for "hypothetical" reserves). The speculative category (Figure 3.1) applies where a geological setting that is likely to contain coal has not yet been explored. State geological surveys working in cooperation with the USGS have been encouraged to adopt this system.

SOURCES OF COAL RESOURCE AND RESERVE INFORMATION

The two primary federal agencies that provide resource and reserve information are the Energy Information Administration in the Department of Energy, and the U.S. Geological Survey in the Department of the Interior.

U.S. Energy Information Administration

The EIA is responsible for maintaining Demonstrated Reserve Base (DRB) data (Box 3.1), the basis for assessing and reporting U.S. coal reserves. The DRB evolved from work performed by the U.S. Bureau of Mines that was published as Information Circulars 8680 and 8693 (USBM, 1975a, 1975b). These circulars contain estimates of DRB tonnage remaining in 1971, reported by county and

---

BOX 3.1
U.S. Demonstrated Reserve Base (DRB)

The DRB is a collective term for the sum of coal in both "measured" and "indicated" resource categories (see Figure 3.1), and includes the following:

- Beds of bituminous coal and anthracite 28 inches or more thick and beds of subbituminous coal 60 inches or more thick that can be surface mined; and
- Thinner and/or deeper beds that presently are being mined or for which there is evidence that they could be mined commercially at this time.

The DRB represents that portion of the identified resources of coal from which reserves are calculated (see Figure 3.2) and is thus a derived value using arbitrary limits and based on limited coal industry data. More recent (2005) numbers for each category except total resources are presented in Table 3.1.

The concept that coal resource and reserve tonnages will sequentially decrease corresponding to greater data reliability and increased confidence in economic recoverability, as portrayed in Figure 3.2, is fundamentally correct. However, there is no justification for the estimates to have more than two significant figures, nor are the sharp boundaries between the different reserve and resource categories realistic. These boundaries will shift up or down (decreasing or increasing tonnage estimates) depending on the availability of new information or particular changes or trends in technology and economics, as well as environmental constraints, transportation availability, and demographic shifts.

coal bed and by sulfur content. The EIA became responsible for maintaining the DRB database in 1977 under the Department of Energy Organization Act of 1977 (P.L. 95-91), which required the EIA to carry out a comprehensive and unbiased data collection program and to disseminate economic and statistical information to represent the adequacy of the resource base to meet near- and long-term demands. Since 1979, EIA has published updates to the DRB by adding additional reserve/resource data from state coal assessments and by depleting the DRB according to the amount of annual coal production. The DRB represents a subset of total national coal resources, because it includes only coal that has been mapped, that meets DRB reliability and minability criteria, and for which the data are publicly available (see Box 3.1 and Figure 3.2). The EIA also reports Estimated Recoverable Reserves (ERR). The ERR is derived from the DRB by applying coal mine recovery and accessibility factors by state to the DRB. The ERR is categorized by state, Btu (British thermal unit) value, sulfur content, and mining type—it is the most widely reported and frequently quoted estimate of U.S. coal reserves.

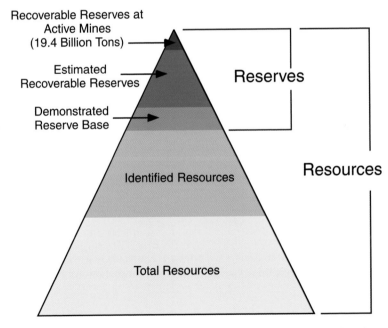

FIGURE 3.2 Triangle depicting U.S. coal resources and reserves, in billion short tons, as of January 1, 1997. The darker shading corresponds to greater relative data reliability. SOURCE: EIA (1999).

The EIA is authorized under federal statutes to collect confidential reserve data from coal companies with active mines. These reserve data are compiled into the Recoverable Reserves at Active Mines (RRAM) database. Although RRAM data are updated annually, they represent only a fraction of the reserves controlled by mining companies. The business complexities of resource holding companies, landowners, lease holders, and production companies make it difficult to collect comprehensive RRAM data, and these data are therefore too limited for mid- and long-range planning. There is no direct relationship between the ERR and the RRAM because they are determined from different data sets.

U.S. Geological Survey (USGS)

The USGS has responsibility for mapping and characterizing the nation's coal resources, in cooperation with agencies that have land and resource management responsibilities (e.g., Bureau of Land Management, Office of Surface Mining Reclamation and Enforcement) and agencies that use USGS resource projections (e.g., EIA). The USGS is in the process of undertaking a systematic inventory of the U.S. coal reserve base to determine how much of the domestic coal endowment is technologically available and currently economic to produce. Assessments of coal quality are a core component of the USGS energy resources program research portfolio (NRC, 1999). The USGS has recently focused its efforts on accumulating data on coal quality for the feed coals and coal combustion products from individual coal-fired power plants, including assessments of elements in coal that can potentially have adverse effects on environmental quality and/or may be slated for regulation. To accomplish its goals, the USGS has a number of programs—in collaboration with other federal and state agencies—that are intended to better characterize the nation's coal endowment.

National Coal Resources Data System. The USGS analyzes coal samples and collects geological data in cooperation with coal-producing states. Thousands of coal samples have been analyzed, and hundreds of thousands of data points describing coal geology, thickness, and depth have been collected. Although this program is still active, it has been scaled down in the past decade because of restricted funding.

Coal Availability Studies. The USGS and state geological surveys have a cooperative program to assess the proportion of identified coal resources that are available to the coal industry for mining. These studies take into account regulatory considerations that restrict mining (e.g., distribution of public lands, streams, or oil and gas wells), as well as technological issues that impede mining such as thin coal seams, mine barriers, seams that are too closely spaced for all to be mined using existing methods, and faulted areas. Thus far, 108 coal availability studies have been completed,[2] and these suggest that less than 50 percent of identified coal resources are available for mining.

---

[2]Presentation to the committee by T. Rohrbacher, U.S. Geological Survey, July 2006.

Coal Recoverability Studies. This program takes the results of coal availability studies and applies engineering criteria to determine minability and recoverability (e.g., Carter et al., 2001). These resource calculations differ from the others described above in that they take in-seam rock partings[3] into account and estimate the percentage of recoverable coal according to anticipated mining methods and coal washability characteristics. A total of 65 areas in 22 coal fields have been analyzed, and these studies suggest that 8 to 89 percent of the identified resources in these coal fields are recoverable and 5 to 25 percent of identified resources may be classified as reserves.[4] Because they are based on site-specific criteria, these studies provide considerably improved estimates compared to the ERR. Ultimately, comprehensive coal recoverability studies would allow cost curves to be generated so that reserve quantities could be determined for different cost levels.

National Coal Resource Assessment. In 1995, the USGS began the National Coal Resource Assessment (NCRA) for major coal beds in selected coal basins by compiling data from adjoining states into a single assessment in GIS (geographic information system) format. The NCRA estimates only the major coal-producing beds and therefore cannot easily be compared with the DRB, which is aggregated for all beds. Some of the NCRA assessments have been updated using coal availability and recoverability criteria to yield basin-wide reserve estimates.

Inventory Studies. The USGS recently initiated a systematic inventory of the U.S. coal reserve base, to determine the subset of in-place resources that is technically and economically recoverable on a basin-wide scale. An initial reserve estimate for the Gillette coalfield of the Powder River Basin is expected in 2007, to be followed by reserve estimates for the entire Powder River Basin by the end of 2008.

Other Sources of Coal Resource and Reserve Information

Although most coal-producing states have geological surveys that collect data on their coal resources, in most cases these organizations lack the personnel and funding for major coal resource and reserve investigations. Most coal resource investigations have been undertaken in cooperation with the USGS, Bureau of Land Management, or the Office of Surface Mining Reclamation and Enforcement. For this reason, state geological surveys typically only evaluate in-place tonnage and do not estimate recoverability—this has been largely left to the USGS and EIA.

Mining companies generate detailed reserve estimates for the coal they control or are interested in obtaining. Companies consider these data to be proprietary, and consequently they are not available for government resource and reserve studies except for the reserve estimates that have to be reported at

---

[3]Rock partings are thin layers of rock within coal seams.
[4]Presentation to the committee by T. Rohrbacher, U.S. Geological Survey, July 2006.

operating mines. Similarly, some states use industry data to prepare coal reserve estimates on unmined reserves for tax purposes, but these data are not publicly available.

The Keystone Coal Industry Manual is a private publication for the coal industry that has been published annually since 1918. It contains descriptions of the coal resources and geology of coal fields for each coal-producing state, describing coal bed geology, stratigraphy, thickness, quality, rank, mining methods, and identified resources (measured, indicated, and inferred). The state sections of the Keystone Coal Industry Manual are updated on an irregular basis, generally by state geological survey geologists.

U.S. COAL RESOURCE AND RESERVE ESTIMATES

The current estimates of total U.S. coal resources and reserves reported by the EIA are shown in Table 3.1. The ERR (Figure 3.2)—approximately 54 percent of the DRB—is calculated based on accessibility factors (by coal-producing region) and recoverability factors at existing mines. ERR and DRB estimates by state and mining method are presented in Appendix D; a subset of these data for the 15 states containing the largest reserves is shown in Table 3.2.

Limitations of Existing Coal Resource and Reserve Estimates

Old and Out-of-Date Data. By definition, the DRB does not represent all of the coal in the ground (EIA, 2006b). It represents coal that has been mapped, that meets DRB reliability and minability criteria, and for which the data either are publicly available or have been provided by companies under confidentiality provisions. The DRB was initiated in the 1970s, and consequently the majority of DRB data were compiled based on the geological knowledge and mining technology available more than 30 years ago. Although the DRB has been updated in 1989, 1993, 1996, and 1999 to incorporate reserve depletion data and limited more recent reserve data (EIA, 1999), the underpinning data remain those of the original 1974 study. The data on Identified Resources and Total Resources

TABLE 3.1 U.S. Coal Resources and Reserves in 2005

Category	Amount (billion short tons)
Recoverable Reserves at Active Mines	19
Estimated Recoverable Reserves	270
Demonstrated Reserve Base	490
Identified Resources (from Averitt, 1975)	1,700
Total Resources (above plus undiscovered resources)	4,000

NOTE: The relationships between these categories are depicted in Figures 3.1 and 3.2.

TABLE 3.2 Estimated Recoverable Reserves and Demonstrated Reserve Base for the 15 States with Largest Reserves, by Mining Method for 2005 (million short tons)

State	Underground Minable Coal		Surface Minable Coal		Total	
	ERR	DRB	ERR	DRB	ERR	DRB
Alabama	508	1,007	2,278	3,198	2,785	4,205
Alaska	2,335	5,423	499	687	2,834	6,110
Colorado	6,015	11,461	3,747	4,762	9,761	16,223
Illinois	27,927	87,919	10,073	16,550	38,000	104,469
Indiana	3,620	8,741	434	742	4,054	9,483
Kentucky	7,411	17,055	7,483	12,965	14,894	30,020
Missouri	689	1,479	3,157	4,510	3,847	5,989
Montana	35,922	70,958	39,021	48,272	74,944	119,230
New Mexico	2,801	6,156	4,188	5,975	6,988	12,131
North Dakota	—	—	6,906	9,053	6,906	9,053
Ohio	7,719	17,546	3,767	5,754	11,486	23,300
Pennsylvania	10,710	23,221	1,044	4,251	11,754	27,472
Texas	—	—	9,534	12,385	9,534	12,385
West Virginia	15,576	29,184	2,382	3,775	17,958	32,960
Wyoming	22,950	42,500	17,657	21,319	40,607	63,819
Other states	8,667	12,226	2,535	3,861	11,202	16,086
U.S. Total	**152,850**	**334,876**	**114,705**	**158,059**	**267,554**	**492,935**

NOTE: Data for all states are shown in Appendix D.

SOURCE: EIA (2006b).

currently published by the EIA are estimates from 1974 that were presented by Averitt (1975) and have not been updated.

Another shortcoming is the restricted use of modern geospatial technology for reserve data management. The current system provides data tables and compiled estimates without supporting map and geographic information. The use of modern GIS-based data management systems would have the advantage of being map-based, reproducible, and updateable as new data become available. In addition, the coal reserve and resource database that supports EIA estimates is out-of-date, and much of the legacy data may be irretrievable due to changes in computer technology in the 30+ years since the DRB was initiated.[5]

Mined Coal Not Included in DRB. Wood et al. (1983) set guidelines for the seam thicknesses and mining depths needed for coal to qualify for the DRB, stipulating that only measured and indicated resources meeting certain conditions could be included (e.g., criteria in Box 3.1). Coal beds are currently being

---

[5]Presentation to the committee by R. Bonskowski, Energy Information Administration, June 2006.

mined that are too deep or too thin to qualify as a part of the DRB under these criteria—some underground mining is being carried out at greater than 2,500-foot depths, and surface (and some underground) coal seams less than 28 inches thick are being mined in the eastern states. Mining technology improvements have resulted in resources not presently included in the DRB becoming economically recoverable and therefore eligible for inclusion in the DRB and the ERR.

Restricted Availability of Industry Data. The EIA, USGS, and state geological surveys typically do not have access to the large amount of private industry exploration and development data that include extensive drilling and active mining information. Although mining company data are occasionally made available for government coal resources studies, federal and state agencies are in general limited to publicly available coal bed related information such as outcrops, road cuts, oil and gas wells, water wells, and maps of abandoned mines. With limited budgets, many coal-producing states have been unable to explore all of their coal resources, resulting in substantial resources being included in the "identified" category when additional information (e.g., more closely spaced data) could result in these resources being confirmed in the DRB and ERR.

Inferred and Undiscovered Resources Ignored. Coal seams are found in a variety of geologic settings and their characteristics, including variability in thickness and continuity, can differ markedly from basin to basin. Therefore, any definition of geological reliability (measured, indicated, and inferred) that is intended for the entire country is not as precise as a system that takes into account the geological differences between regions and between coals of different geological ages. Although Wood et al. (1983) permit practitioners to specify customized dimensions for reliability circles to reflect the variability of coal deposits, most states use the recommended ¼-, ¾-, and 3-mile data spacing (for measured, indicated, and inferred, respectively) to facilitate comparisons with other estimates. This means that reserves existing ¾ mile to 3 miles from a point of coal measurement (e.g., a drill hole or outcrop) are classified as "inferred," and all coal existing beyond a 3-mile radius falls into the "undiscovered" category. As a result, a large amount of coal in the "inferred" category is not in the DRB and is not included in ERR calculations.

Alaska provides an example of potential coal reserves not accounted for by EIA statistics. The most recent comprehensive state coal resource assessment indicates that total hypothetical coal resources in Alaska exceed 5.5 trillion short tons (Merritt and Hawley, 1986). By comparison, the EIA-USGS estimate of total U.S. resources, including hypothetical measures, is 4 trillion tons. Alaska accounts for only 1 billion tons in the 2004 DRB estimate, even though state experts consider that coal reserves in Alaska may possibly surpass the total coal resources in the lower 48 states.

Coal Quality Issues. The uncertainties concerning resource and reserve estimates also apply to the grade or quality of the coal that will be mined in the future. At present, we lack methods to project spatial variations of many important coal quality parameters beyond the immediate areas of sampling (mostly drill samples). Almost certainly, coals mined in the future will be lower quality because current mining practices result in higher-quality coal being mined first,[6] leaving behind lower-quality material (e.g., with higher ash yield, higher sulfur, and/or higher concentrations of potentially harmful elements). The consequences of relying on poorer-quality coal for the future include (1) higher mining costs (e.g., the need for increased tonnage to generate an equivalent amount of energy, greater abrasion of mining equipment); (2) transportation challenges (e.g., the need to transport increased tonnage for an equivalent amount of energy); (3) beneficiation challenges (e.g., the need to reduce ash yield to acceptable levels, the creation of more waste); (4) pollution control challenges (e.g., capturing higher concentrations of particulates, sulfur, and trace elements; dealing with increased waste disposal); and (5) environmental and health challenges. Improving the ability to forecast coal quality will assist with mitigating the economic, technological, environmental, and health impacts that may result from the lower quality of the coal that is anticipated to be mined in the future.

INTERNATIONAL COAL RESOURCE ASSESSMENTS

The World Energy Council (WEC) publishes, on a triennial schedule, a *Survey of Energy Resources*, the most recent of which is the twentieth edition (WEC, 2004). This survey includes fossil fuels (coal, oil, and natural gas), uranium and nuclear fuel, and renewable resources. Where relevant, tables are published of fossil fuel resources and reserves, with data derived from member countries of the WEC and from non-WEC sources. The WEC data tables for coal are widely accepted, used, and quoted by numerous agencies and entities (e.g., IEA, 2004; BP, 2006; EIA, 2006f). Collecting reliable and comprehensive data on a worldwide basis, from more than 75 countries, presents a significant challenge for the authors of the publication—in particular, they note that resource and reserve definitions can differ widely among countries (WEC, 2004). In an attempt to improve the data on resources and reserves of fossil fuels and uranium, the WEC has been coordinating with the United Nations Economic Commission for Europe (UNECE) seeking to adopt a uniform set of definitions; however, uniform definitions were not in place for the 2004 edition. International data for proved recoverable reserves presented by WEC (2004) are listed in Appendix D.

The 10 countries that reported the largest quantity of proved recoverable reserves (Figure 3.3) have, in aggregate, 92 percent of the world's proved recoverable reserves. The top three countries—the United States, the Russian Federation, and China—contain 57 percent of proved recoverable reserves. China, which produced 40 percent more coal in 2002 than the United States, reported proved

---

[6]This practice is known as highgrading.

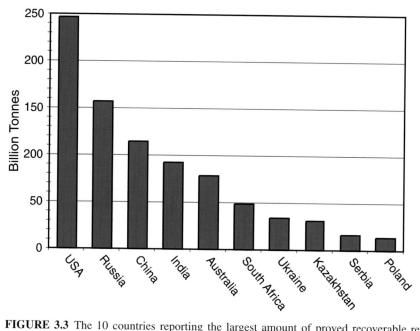

FIGURE 3.3 The 10 countries reporting the largest amount of proved recoverable reserves in 2002. SOURCE: Data from WEC (2004).

recoverable reserves that were only 46 percent those of the United States—an anomalously low value (WEC, 2004).

It is possible to undertake the academic exercise of dividing the worldwide proved recoverable reserves by the total world coal production for the same year, to obtain about 188 years of production. Although correct mathematically, this number is of little value because it suffers from the same inconsistencies and deficiencies in input parameters as the equivalent calculation for the United States. Like the United States, the world has vast amounts of coal resources, and like the United States, a clear picture of global coal reserves is difficult to ascertain. In part, this is due to strategic concerns about revealing information on domestic energy resources, absence of government recognition of the importance of such information, the lack of trained personnel or funding to carry out such studies, and differences in methodology and terminology.

FINDINGS AND RECOMMENDATION—COAL RESOURCE, RESERVE, AND QUALITY ASSESSMENTS

Federal policy makers require sound coal reserve data to formulate coherent national energy policies. Accurate and complete estimates of national reserves are needed to determine whether coal can continue to supply national electrical power needs and to determine whether coal has the potential to replace other energy sources that may become less reliable or less secure.

• The United States is endowed with a vast amount of coal. Despite significant uncertainties in generating reliable estimates of the nation's coal resources and reserves, there are sufficient economically minable reserves to meet anticipated needs through 2030. Further into the future, there is probably sufficient coal to meet the nation's needs for more than 100 years at current rates of consumption. However, it is not possible to confirm the often-quoted suggestion that there is a sufficient supply of coal for the next 250 years. A combination of increased rates of production and more detailed reserve analyses that take into account location, quality, recoverability, and transportation issues may substantially reduce the estimated number of years of supply. Because there are no statistical measures to reflect the uncertainty of the nation's estimated recoverable reserves, future policy will continue to be developed in the absence of accurate estimates until more detailed reserve analyses—which take into account the full suite of geographical, geological, economic, legal, and environmental characteristics—are completed.

• The Demonstrated Reserve Base (DRB) and the Estimated Recoverable Reserves (ERR), the most cited estimates for coal resources and reserves, are based on methods for estimating resources and reserves that have not been reviewed or revised since their inception in 1974. Much of the input data for the DRB and ERR are also from the early 1970s. These methods and data are inadequate for informed decision making. New data collection, in conjunction with modern mapping and database technologies that have been proven to be effective in limited areas, could significantly improve the current system of determining the DRB and ERR.

• Coal quality is an important parameter that significantly affects the cost of coal mining, beneficiation, transportation, utilization, and waste disposal, as well as the coal's sale value. Coal quality also has substantial impacts on the environment and human health. The USGS coal quality database is largely of only historic value because relatively few coal quality data have been generated in recent years.

Recommendation: A coordinated federal-state-industry initiative to determine the magnitude and characteristics of the nation's recoverable coal reserves, using modern mapping, coal characterization, and database technologies, should be instituted with the goal of providing policy makers with a comprehensive accounting of national coal reserves within 10 years.

The U.S. Geological Survey already undertakes limited programs that apply modern methods to basin-scale coal reserve and quality assessments. The USGS also has the experience of working with states to develop modern protocols and standards for geological mapping at a national scale through its coordinating role in the National Cooperative Geologic Mapping Program. The USGS should be funded to work with states, the coal industry, and other federal agencies to quantify and characterize the nation's coal reserves. Recognizing the urgency of this requirement, the committee stipulated that this comprehensive accounting should be completed within 10 years, although it accepts that the exact time frame may be shortened by the lead agency on approval of the project. The committee estimates that this will require additional funding of approximately $10 million per year.

4

Coal Mining and Processing

Although the United States has the vast coal resource described in the previous chapter, perhaps as much as 4 trillion tons, the key issue for policy makers is the amount of coal that is economically recoverable. This is not a fixed quantity, but depends on the geological resource, the market price, and the cost of mining. The particular characteristics of the coal mining industry create unique challenges as it endeavors to provide coal to the market at a competitive price, continually improve miner safety and health, and meet environmental and community requirements. This chapter provides a brief description of the characteristics of the coal industry in the United States, presents an overview of coal extraction and processing methods, and discusses the major issues associated with coal mining and processing. A summary of current research activities supported by state and federal agencies provides the context for recommendations for essential future research.

COAL MINING INDUSTRY IN THE UNITED STATES

The U.S. coal industry serves a vital role in the nation's economy by producing fuel for more than half of its electricity. Despite the industry's importance, industry financial data for 2005—the strongest year for the coal industry in recent years—shows that it is a relatively small industry with revenues totaling $20 billion to $25 billion and net income between $1 billion and $2 billion. To put that in perspective, the entire industry taken collectively would rank as about the one-hundredth largest company (in terms of either revenue or net income) on the 2005 "Fortune 500" list, and it is less than 10 percent the size of Wal-Mart.

The U.S. coal industry has undergone a remarkable transformation over the last three decades. During this time, coal production has doubled, while the number of active miners has been halved and the number of mines has dropped by a factor of three (Figure 4.1). This has resulted in the concentration of production in a smaller number of larger mines. The 100 largest mines in the country produced 805 million tons of coal in 2004 (72.5 percent of total U.S. production), while employing about 45 percent of the mining workforce (an average of 310 miners per mine) (EIA, 2005d; NMA, 2006a). The remaining approximately 1,300 mines produced 27.5 percent of the nation's coal while employing 55 percent of the workforce (about 30 miners per mine). Nearly 70 percent of U.S. coal mines, many of which are comparatively small operations, are in Kentucky, West Virginia, and Pennsylvania.

Since the 1970s, there has been a continuous increase in the proportion of coal produced by the western states. At present, states west of the Mississippi account for more than 55 percent of total tons produced (Figure 4.2). Wyoming alone accounts for almost 36 percent of national coal production tonnage (Table 4.1).

Considerable data are compiled on the basis of the union or non-union status of mines throughout the coal industry by the Energy Information Administration (EIA). At present, some 27.5 percent of the total coal mining workforce consists

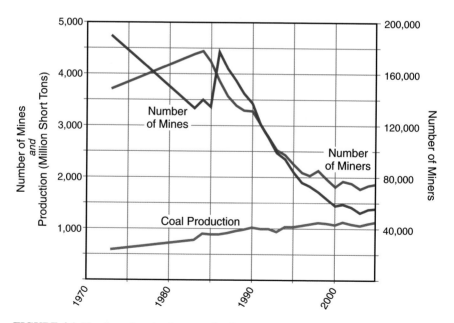

FIGURE 4.1 Number of coal mines, production tonnage, and miner employment in the U.S. coal industry. Note that the left axis scale represents two parameters, production tonnage and number of mines. SOURCE: Based on data in NMA (2006b).

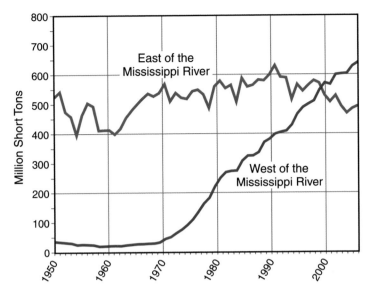

FIGURE 4.2 Domestic coal production during the last half century across the United States. SOURCE: EIA (2006a).

of union members, working at 147 mines (EIA, 2005d). The United Mine Workers of America (UMWA) represents the largest number of workers (16.3 percent of the workforce) and the largest number of mines (131) of any of the unions. The average number of workers at a union mine is 140.

COAL MINING AND PROCESSING METHODS

A modern coal mine is a highly mechanized industrial plant that has to meet strict standards of engineering design and operation. The size, power, strength, monitoring and control features, and automation of mining equipment dwarf

TABLE 4.1 U.S. Top Coal Producing States in 2005

State	Production in 2005 (thousands short tons)	Percentage of Total U.S. Production Tonnage
Wyoming	404,319	36
West Virginia	153,650	14
Kentucky	119,734	10
Pennsylvania	67,494	6
Remaining coal-producing states	386,301	34
U.S. Total	**1,131,498**	**100**

SOURCE: EIA (2006b).

those of even a few decades ago. Coal mines require substantial capital invest-ment in both permanent structures and depreciable mining equipment, exceeding $75 per annual ton of capacity in large underground coal mines and $30 or more per annual ton for large surface coal mines.

The overall coal mining process consists of several sequential stages: (1) exploration of a potentially economic coal seam to assess minable reserves, environmental issues, marketable reserves, potential markets, and permitting risks; (2) analysis and selection of a mining plan; (3) securing the markets; (4) developing the mine; (5) extracting the coal; (6) processing the coal if necessary; and (7) decommissioning the mine and releasing the property for post-mining use. The two essential requirements that must be fulfilled before a prospective coal mine can enter the development stage are confirmation that there are sufficient minable reserves of adequate quality with no unacceptable environmental or permitting risks, and confirmation of an assured or contracted market for a substantial fraction of the coal that will be mined.

Coal Mining

Coal seams can be mined by surface or underground methods (Figure 4.3), with the choice of mining method dictated by both technical and economic fac-tors. The most important technical factors are the thickness of the coal seam, the depth of the coal seam, the inclination of the seam, and the surface topography.

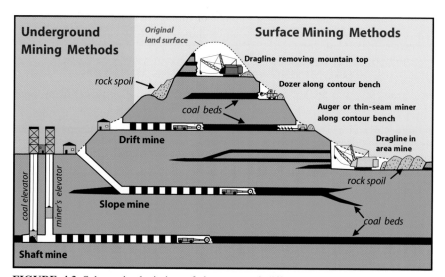

FIGURE 4.3 Schematic depiction of the range of different surface and underground types of coal mining, illustrating types of access to coal deposits and mining terminology. SOURCE: KGS (2006).

Each of these technical factors can set limiting conditions when considering the economic recoverability for a given coal, mining method, and market. The important economic parameters are the relative costs of mining coal by surface and underground methods including costs associated with any site-specific land use constraints, the cost of removing the material above the coal seam in the surface method, and the price of coal. The price for any particular coal is related directly to coal quality (Box 4.1 and Table 4.2). Because more than 90 percent of the coal mined in the United States is used in power plants to generate electricity using steam turbines, the price for steam coal is dependent primarily on its heating value and sulfur content.

Relatively shallow coal deposits are generally extracted by surface mining, and deeper deposits are extracted by underground mining (more detailed descriptions of surface and underground mining processes can be found in Appendix E). There are also situations in which a seam is mined by surface methods first, and then if adequate reserves are still available, the mine is developed for underground extraction. Where remaining reserves are limited, other methods of mining—such as auger mining or highwall mining—may be used.

Surface mining has many advantages compared to underground mining. In general, coal recovery is very high (85 to 90+ percent), compared to 40 to 70 percent in underground mines. The productivity of surface mines is generally higher than that of underground mines (Figure 4.4), and health and safety statistics for surface mining are also generally better than those of underground mining. Surface-mined coal from the Powder River Basin is usually simply sized and screened in preparation for market, whereas underground-mined coal and surface-mined coal from the Interior and Appalachian basins often requires a greater amount of processing (see below) to improve its marketability. The cost per ton of mining coal by surface methods is generally lower than that by underground methods.

In the United States, in addition to a continuous growth in coal production since the 1960s, there has been a dramatic shift in production from underground mining to surface mining (Figure 1.6). In the Powder River Basin (PRB), where deposits of coal more than 100 feet thick occur close to the surface, individual surface mines can produce more than 90 million tons each year. Underground coal mining is more common east of the Mississippi River, particularly in Appalachia. Some of the largest underground coal mines, each producing around 10 million tons annually, are located in Pennsylvania and West Virginia. The largest underground mining complex in the United States produces about 20 million tons per year.

Bituminous coals in the eastern and central United States are mined by both surface and underground mining methods. Anthracite coal is mined exclusively in northeastern Pennsylvania, also by both underground and surface mining methods. Lignite and subbituminous coal production is centered in a small number of large mines (Table 4.3). Subbituminous coal and lignite comprise about 50 percent of U.S. coal production by tonnage, but because of their lower heating values, only about 40 percent by heating value. The distinction between tonnage and energy content is particularly important when considering possible shifts in coal supply and demand by producing and consuming regions. Subbituminous coals are produced almost exclusively in the PRB of Wyoming and Montana. EIA's forecasts of increased coal production over the next three decades (see

---

BOX 4.1
Coal Rank and Quality

Although the term coal refers to any readily combustible rock containing more than 50 percent by weight of organic matter, coals differ considerably in their physical and chemical characteristics (Table 4.2) and these differences have pronounced impacts on their value and use. Coals in the United States are classified by "rank," a method of distinguishing coals on the basis of their fixed carbon content, volatile matter content, heating value, and agglomerating characteristics.[1] Coal rank is defined as "the degree of metamorphism, or progressive alteration, in the natural series from lignite to anthracite. Higher-rank coal is classified according to the fixed carbon on a dry basis, lower-rank coal, according to Btu [heating value] on a moist basis" (AGI, 1997).[2] Differences in sedimentary depositional environments and differences in the geological history of the coal strata result in differences in mineral matter content and composition, as well as differing concentrations of some of the important minor elements (e.g., sulfur, chlorine). These differing characteristics impact coal utilization in both the electricity generation and metallurgical markets.

The term *coal quality* is used to distinguish the range of different commercial steam coals that are produced directly by mining or are produced by coal cleaning.[3] Generally, coal quality for steam coals (i.e., coal used for electricity generation) refers to differences in heating value and sulfur content (Table 4.2), although other characteristics such as grindability or ash fusion characteristics are also specified in coal sale agreements. While not as obvious as the impact of sulfur content on environmental emissions, differences in the moisture content and heating values among different coal types affect CO_2 emissions upon combustion, with higher-rank bituminous coals producing 7 to 14 percent lower emissions than subbituminous coals on a net calorific value basis (Winschel, 1990).

---

[1]This classification is described in American Society for Testing and Materials (ASTM) standard D388-77. Standards that are broadly similar, but differ in detail, are used by the international coal trade and some coal mining countries.

[2]Both high- and low-rank carbon content calculations are reported on a mineral-matter-free basis.

[3]Factors considered in judging a coal's quality are based on, but not limited to, heat value; content of moisture, ash, fixed carbon, phosphate, silica, sulfur, major, minor, and trace elements; coking and petrologic properties; and organic constituents considered both individually and in groups. The individual importance of these factors varies according to the intended use of the coal.

TABLE 4.2 Typical Values for Characteristics of Major Commercial Steam Coals in the United States

Rank	Northern Appalachia High Volatile A Bituminous	Central Appalachia High Volatile B Bituminous	Illinois Basin High Volatile C Bituminous	Powder River Basin (Wyoming and Montana) Subbituminous C	Lignite (Texas and North Dakota) Lignite A
Moisture (wt%) as received (AR)	7.0	7.5	12.5	29.8	32.0
Ash (wt%), AR	7.6	8.5	11.0	4.8	10
HHV[a] (Btu/lb), AR	13,100	12,650	11,300	8,500	7,200
Sulfur (wt%), dry	2.5	1.0	2.7	0.5	1
Chlorine, wt%, dry	0.1	0.1	0.2	0.01	0.01

NOTE: Data assembled from numerous sources.
[a]Higher Heating Value.

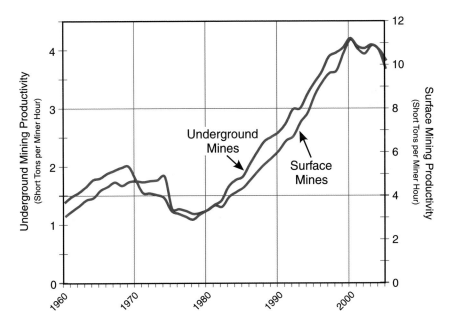

FIGURE 4.4 Productivity trends for surface and underground coal mines, illustrating the dramatic productivity increases over the past three decades. SOURCE: EIA (2006a).

TABLE 4.3 Coal Production by Rank for 2005

	Production (thousand short tons)		No. of Mines	Production per Mine (thousand short tons)	Total Energy Content, (quadrillion Btu)	
Lignite	83,942	(7%)	21	3,997	1.2	(5%)
Subbituminous	474,675	(42%)	30	15,823	8.1	(35%)
Bituminous	571,177	(51%)	1,293	442	13.7	(60%)
Anthracite	1,704	(0.2%)	71	24	0.1	(0.4%)
Total	**1,131,498**		**1,415**	**800**	**23.0**	

NOTE: Because of their lower energy contents, lignite and subbituminous coals represent a smaller percentage of coal production in the United States on an energy basis (~40 %) than on a tonnage basis (~50 %).
SOURCE: EIA (2006b).

Chapter 2) rely heavily on increased production from this region for consumption east of the Mississippi River. Because it takes about 50 percent more subbituminous coal (on a tonnage basis) to replace a ton of bituminous coal in electricity generation,[1] this has significant implications for transportation infrastructure and power plant design and capacity.

When used for electricity generation, coal from the PRB generally produces more CO_2 per kilowatt-hour than the bituminous coal mined in the east. Combustion of subbituminous coal from the PRB produces about 226 pounds of CO_2 for every million Btu (British thermal units) of heat generated (on a net calorific basis), compared with about 211 pounds for the bituminous coal mined in the East (Winschel, 1990). Another possible constraint on the use of coal from the southern PRB might be future air quality regulation of coarse particulates, although Environmental Protection Agency's (EPA's) proposal to exempt mining and agricultural operations in its update of particulate standards (EPA, 2006c) may remove or defer this potential constraint.

An increasingly important by-product of U.S. coal production is coal mine methane recovered during or prior to mining, in addition to coal bed methane produced independently of mining. Captured methane may be used as a fuel source at the mine or, where feasible, distributed in natural gas pipelines. In recent years, coal bed methane production has increased and now comprises about 8 percent of the U.S. natural gas supply. An ancillary benefit of recovering coal mine methane is reduced atmospheric methane emissions, because methane is a potent greenhouse gas.

---

[1] In addition to the difference in heating value (i.e., Btu/lb), electricity generating units fueled with subbituminous and lignite coals tend to operate at lower efficiency (higher heat rate) than units fueled with bituminous coal. This can lead to differences in generating capacity when using different coals.

Coal Processing

"Raw" or "run-of-mine" coal can be processed[2] using physical separation methods to remove unwanted mineral matter to produce a "clean" coal. Processing adds value in several ways:

- Removal of the mineral matter (or "ash"),[3] which is largely noncombustible and may constitute up to 65 percent of the raw coal, increases the heating value of the coal on a mass basis.[4] Although some combustible material is lost as part of the cleaning process, the removal of unwanted material reduces the mass and volume of coal for a given heating value thereby reducing shipping costs as well as minimizing coal handling and ash management costs for the end user.
- Processing allows greater control over the "quality" of the coal—principally ash and moisture—which improves its consistency for end users, such as electricity generators or coke manufacturers. Improved and consistent quality increases the efficiency and availability of steam boilers and is particularly important for the quality of metallurgical coke.
- Physical processing (see Appendix E) can, to some extent, reduce sulfur and trace element contents, particularly on a heating value basis. However, generally coal cleaning is not practiced primarily for this purpose except for the metallurgical coal market.

The decision whether or not to process a particular raw coal depends on the coal and its intended market. The subbituminous coal of the Powder River Basin is almost always shipped to market raw because it has inherently low ash content and poor "washability,"[5] and the region has low water availability—a critical requirement for conventional coal beneficiation.

Most coal preparation plants are in the eastern states, with more than 80 percent of the plants and almost 80 percent of capacity located in West Virginia, Virginia, Pennsylvania, Ohio, and Kentucky (Fiscor, 2005; see Appendix E). The 11 coal preparation plants in the western states are located at bituminous

---

[2]The terms "coal preparation," "cleaning," "washing," "processing," and "beneficiation" are used synonymously to refer to physical enrichment of the combustible portion of the coal by selective removal of the noncombustible components, principally mineral matter and water.

[3]It is customary to refer to the mineral content of a coal as "ash," and it is usually reported as such in coal quality descriptions. Ash content is determined by combusting the coal in air and converting the inorganic elements to their oxides.

[4]Commercial coal characteristics, such as heating value, ash, moisture, sulfur, etc., are determined in the United States according to standards established by the American Society for Testing and Materials (ASTM), and are usually denominated in English units (e.g., Btu/lb for heating value on a mass basis).

[5]The term "washability" is used to describe the ease with which mineral matter can be separated from the coal, and depends on the degree of incorporation of the mineral matter in the coal's organic matrix and its specific gravity relative to the coal.

coal mines. Wyoming—the largest coal-producing state in the country—has no coal preparation plants, largely because the subbituminous Powder River Basin coal is low in ash. However, PRB coal has a high level of inherent moisture (~30 percent), which has stimulated some interest in thermal dewatering of the coal to increase its heating value and reduce its transportation costs (see Box 4.2).

MAJOR COAL MINING AND PROCESSING ISSUES

The conditions that will be encountered in future coal mines will undoubtedly be different from those of today—the more easily mined coal has already been extracted. As shallower coal is depleted and seams with greater amounts of over-burden[6] are mined, surface mining will become more expensive because stripping ratios[7] will increase and multiple benches[8] will be needed. This will increase the number of unit operations and the associated cost. In underground mining, the mines of the future will have to access seams that are deeper, thinner, or thicker, often with higher gas content and potentially with greater ground-control issues. Overlying or underlying seams may have been mined, or it may be necessary to mine multiple seams simultaneously to meet increased production requirements. These more difficult mining situations will have an effect on economic decisions related to mine profitability, on the health and safety of mine workers as they encounter more challenging or different mining requirements, on the technical ability to mine, and on the management of waste materials generated by mining.

Small mines (i.e., with annual production less than 2 million tons), which currently produce more than 25 percent of coal in the United States, play an important role in ensuring adequate coal supply because they have historically started and ceased production as demand and prices fluctuate. They also play a critical role in fully utilizing national coal resources, particularly those resources that may not be mined by larger operations.[9] The population of small coal mines has been decreasing and is projected to continue to decline. However, small mines will continue to exist, and the technical and societal issues they will face in the future should be considered in agency plans.

Miner Health and Safety

Although statistics show substantially improved health and safety conditions in mines in recent years—with continuous decreases in both the incidence and

---

[6]Overburden is the overlying rock and soil that must be removed to gain access to a coal seam to be mined.

[7]The ratio of overburden thickness to coal thickness; may also be measured by weight or volume.

[8]If overburden exceeds a certain thickness, equipment size requires that the overburden is removed in more than one pass leaving a flat bench for equipment access between each pass.

[9]Larger mining companies will only mine when sufficient coal reserves are available to sustain the considerable capital investment required for advanced mining equipment and technologies.

BOX 4.2
Upgrading of Low-Rank Coals

Combined reserves of subbituminous coal and lignite (known as "brown coal" in the international coal trade) make up approximately one-half of the world coal reserves and about one-half of the coal resources of the United States. These coals are rarely processed before shipment or use. However, the oxygen and moisture contents of low-rank coals are greater than those of bituminous coals. This reduces the heating value of the coal as mined, which increases the transportation cost on a heating value basis and reduces the thermal efficiency of the steam boilers that use these coals. Most lignite mined in the United States is used in minemouth plants. Subbituminous coals, however, are generally transported considerable distances, so their high moisture content and low heating value add to the effective transportation cost and environmental impact. One way to offset these disadvantages is to dry the coal before transportation or utilization.

Numerous processes for drying low-rank coals to upgrade them have been proposed, demonstrated (e.g., Great River Energy Lignite Drying Process; Bullinger et al., 2006), and in a few cases, commercialized (e.g., K-Fuels Process; Kowalski, 2005). The characteristics of dried low-rank coal—it is friable, has a tendency to spontaneously heat, and readily reabsorbs moisture—constitute major obstacles that must be overcome to produce a saleable, transportable, dry coal product.

---

the severity of diseases, disasters, fatal accidents, and nonfatal accidents—the health and safety of miners remain a major concern for government, industry, and labor. As the coal mine disasters in early 2006 demonstrated (MSTTC, 2006), the safe operation of mines remains a major challenge—there needs to be constant monitoring and control of health and safety threats as well as continuous safety training and improvements in operating practices.

Past experience has shown that changes in mining operations or practices (e.g., introduction of new equipment and systems, mining of virgin areas, infusion of new workers) all have the potential to create a more hazardous environment. Similarly, experience has shown that adequate engineering controls and a knowledgeable workforce are the prerequisites for a safer work environment. Continued health and safety research is needed to identify new hazards and hazard sources as well as to improve the engineering controls for existing hazards, particularly through the development of reliable monitoring and intelligent control systems. The likelihood of deeper mines in the future means that there has to be increased attention to methane control (including methane capture before, during, or after mining), dust control, ignition sources, fires, and explosions. Recent disasters have shown that there are major knowledge gaps and technology needs in the areas of escape and survival, and emergency preparedness and rescue, emphasizing the need for research to develop systematic and comprehensive risk management protocols that can then be applied to individual mines (MSTTC, 2006). In addition, since powered haulage and machinery now have the dubious honor of surpassing ground-control incidents (e.g., roof falls) as the major source of accidents in mines, there is a clear need for better understanding of the hazards that are likely to be encountered as equipment sizes and operational conditions change. New technology for better geological characterization as part of mine planning, better monitoring sensors, and increased remote control and automation of mining equipment have the potential to decrease miners' exposure to hazardous conditions. Mining of deeper seams, multiple seams, thicker seams, and seams underlying or overlying mined-out seams will all require a greater understanding of strata control aspects. The risks associated with mining coal seams adjacent to previously mined-out seams, with their actual or potential void spaces, emphasize the need for accurate, comprehensive, and readily available mine maps showing the distribution of older mine workings.

Exploration and Mining

Adequate information on the nature and characteristics of a coal seam prior to mining is vitally important for safe and efficient mine operations. Any unex-pected anomalies in a coal seam and associated strata, such as sand channels, washouts, faults, and roof instability, can substantially impair mining productivity and create safety hazards. Research to devise improved techniques for imaging coal seams and associated strata, including coal quality parameters, prior to and during mining is needed for mine planning, permitting, and operations (NRC, 2002b). The Martin County coal impoundment failure (NRC, 2002a) and the Quecreek inundation (PDEP, 2002) illustrated the need for increased accuracy and reliability of the geological information used for mine and mine facilities planning.

At present, the drilling and blasting required for overburden fragmentation is the major limitation on increased surface mining productivity, and the devel-opment of improved rock fragmentation practices is an essential requirement for increased surface mine production. For underground mining, the increased use of longwall mining (see Appendix E) offers the greatest potential for higher productivity. While deeper reserves will be ideal for the increased application of longwalls, a number of limitations to the current production potential of long-walls, in particular the need for better roof support and improved coal haulage systems, must be overcome. Other areas in which the development of advanced technologies offers considerable potential for increased productivity are the con-tinuous monitoring of produced coal and the development of improved remote control, automatic control, and autonomous systems.

Selective mining and blending are two practices that have been advocated to decrease the handling of unnecessary waste during mining and processing, and to increase the utilization of all coals for a range of purposes. While the bulk mining methods of the present do not generally favor such practices, these areas need to be explored to determine whether they offer potential for increased mining productivity.

Even when exploration shows promise of an economically minable deposit, the elapsed time from first investment in planning until a mine enters full production, after passing through the permitting, construction, and marketing processes, can take anywhere from seven to fifteen years for a large operation. Although smaller operations in established coal mining districts may take less time, two to five years is normal even in these situations. This time delay can significantly impact the economic feasibility of opening a new mine and will have to be minimized if the higher production scenarios for the future are to be achieved.

Coal Processing

Research to improve coal processing has considerable potential to optimize the use of the nation's coal resources by increasing production and productivity and improving environmental protection. The effect of improved marginal economics by increasing coal recovery can be significant—in one case, a study showed that a 1 percent increase in recovery of coal could increase profits by 25 percent.[10] This could have the effect of making an uneconomic reserve available for production, thus extending the nation's economic reserve base.

- There are large volumes of western bituminous coals with high ash content (>20 percent) that cannot be cleaned by conventional methods because there are insufficient water resources where the coal is mined to permit conventional wet coal processing. Also, a considerable amount of surface-mined subbituminous coal is lost because of out-of-seam dilution with mineral matter—an annual loss of as much as 10 million tons was reported for the Arch Coal Black Thunder Mine (which produces about 80 million tons per year).[11] The development of dry coal beneficiation processes designed for these coals could greatly increase recoverable reserves.
- Improved coal processing also offers the potential to minimize existing environmental problems and potential future issues. There is approximately 2 billion tons of waste fine coal in "gob" piles resulting from past processing techniques in eastern bituminous coal fields—Pennsylvania alone reports more than 250 million tons of waste coal (McGinty, 2004). Many of these piles are environmental liabilities being dealt with under the federal Abandoned Mine Land reclamation program, but a growing number are being viewed as potential opportunities for utilization. Pennsylvania has 14 sites at which circulating

---

[10]Presentation to the committee by G.H. Luttrell, Virginia Tech, June 2006.
[11]Presentation to the committee by P. Bethell, Arch Coal, Inc., September 2006.

fluidized bed boilers are operating on waste coal for electricity or process steam generation.

• Improved coal processing also offers potential for responding to future environmental requirements. For example, the development of new or modified flotation processes permit fuel oil to be replaced as a froth flotation collector if it is prohibited because of disposal concerns.

There are two technical areas where the development of improved coal processing technologies offers the greatest potential to increase resource recovery (Peterson et al., 2001; NRC, 2002b; CAST, 2003):

• The use of improved information technology, perhaps in conjunction with improved online analysis capabilities, to optimize the performance and efficiency of existing unit operations; and
• The development and deployment of better materials with which to construct vessels, separation devices, and conduits.

Mining and the Environment

The primary needs for research in the broad environmental area are to support the regulation of existing and future mining operations and to mitigate the effects of past mining practices.

Existing Mine Operations. There is still an incomplete understanding of how strata behave after coal is extracted from both surface and underground mines, and the hydrologic consequences of mining are not fully understood. For surface mining, the properties of the altered subsurface—particularly the leaching and permeability characteristics—are likely to be different compared to those existing prior to mining. For underground mining, the collapse of strata above a coal seam into the mined void can propagate all the way to the surface, damaging buildings and disrupting the quantity and quality of surface and subsurface water flows.

Disposal of mine waste can be a significant problem, particularly where the coal has to be cleaned before shipment (e.g., see NRC, 1975; 1981; 2002a; 2006). There is a need for enhanced understanding of the physical and chemical behavior of spoil stored in valleys or waste—from coal combustion or coal preparation plants—that is disposed in surface or underground mines. Waste management is a major problem where land either is not available or is more valuable for other productive uses. Increased research to develop productive uses of mine waste offers considerable potential to reduce waste disposal issues.

Mine Decommissioning and Closure. Federal regulations for decommissioning and closure of mining operations are administered by the Mine Safety and Health Administration (MSHA), the Office of Surface Mining Reclamation and Enforcement (OSM), and the EPA—in some cases state and local governments have additional requirements. The major decommissioning and closure activities are (1) sealing of all access to underground mine areas, (2) removal of all surface facilities, and (3) reclamation of surface mine areas (generally carried out concurrently with mining operations) and the surface areas of underground mines. Underground and surface coal mines present different challenges for decommissioning and closure. The critical factors in underground mining are the effects of subsidence and hydrology, both of which require continued monitoring and control. For surface mines, the critical factors relate to drainage and treatment of water and to erosion and sedimentation of the slopes, the waste and spoil banks, and the final pit. Continued use of the surface mine infrastructure (e.g., roads, buildings, utilities) depends to a large extent on the post-mining requirements described in the mining plan. A mining plan that is well integrated with a community master plan can result in optimum post-mining use of this infrastructure.

Abandoned Mined Lands. A range of environmental issues (e.g., subsidence, fires, acid mine drainage, waste disposal sites, derelict lands) associated with abandoned mined land (AML) continues to cause major concerns and threats to the health, safety, and general welfare of communities. This problem is particularly acute in the older coal mining districts of the eastern United States, specifically in the Appalachian hill country. Although mine closure today is a rigorously regulated process requiring detailed technical and financial analysis during the planning and operation stages for a mine—and ensuring financial and legal responsibility for post-mining closure—the nation continues to grapple with the effects of past mining practices. Additional research is required to develop and demonstrate more effective and sustainable solutions to the problems of acid mine drainage, mine fires, and the utilization of waste piles from AML sites.

COAL MINING AND PROCESSING R&D PROGRAMS

Coal mining research and development (R&D) are carried out by a range of organizations and entities—federal government agencies, state government agencies, equipment manufacturers, academic institutions, and industry. In general, the scope of and motivation for research are determined by the relevance and potential impact of the problems that need to be dealt with by these various stakeholders. Industry participants in mining research include individual companies and mining company associations.

Research Programs in Federal Agencies

While the federal government continues to have extensive involvement in the regulation of the coal mining industry, its support for mining research has decreased substantially over the past 10 years. At present, federal research is focused primarily on health and safety. Some research is being done on environmental issues, but support for research aimed at advanced mining technologies and practices has progressively declined since the closure of the U.S. Bureau of Mines (USBM) in 1995 and is now essentially eliminated (see Appendix C).

Engineering and Technology Development. The now-defunct Mining Industry of the Future (IOF) program, administered by the Department of Energy's (DOE's) Office of Energy Efficiency and Renewable Energy, supported engineering and technology development with a focus on improving the energy efficiency, resource utilization, and competitiveness of the mining industry. Although not exclusively focused on extraction or on coal mining, many of the program outputs were applicable to the extraction phase of the coal fuel cycle. At present, coal extraction receives no support from the DOE-Office of Fossil Energy (FE) Coal R&D program, which is focused primarily on utilization aspects (see Chapter 6).

Relatively little is being done by the federal government to address coal preparation issues. The Mining IOF program funded some work that was relevant to minerals separation (as part of a much broader program and not exclusively coal related), but new funding for this program (never more than $5 million per year) has been terminated and the program is being closed out. DOE-FE had a solid fuels program, although it tended to fund more advanced work—such as chemical coal cleaning—than processes related to conventional coal preparation. However, there has been no administration request for funding for this area in recent years, and the program is essentially defunct. Some research programs addressing a variety of mineral separation issues (i.e., not exclusive to coal) have been funded at the federal level through small direct congressional appropriations.

There is a low level of support for fundamental research in the earth sciences and engineering disciplines (geosciences, material sciences, rock mechanics, etc.) by the National Science Foundation that has potential applications in the development of improved technologies for the coal industry.

Health and Safety. The National Institute for Occupational Safety and Health (NIOSH) Mining Program is the principal focus for mining health and safety research (NRC, 2007b), consolidated at the Pittsburgh and Spokane research centers. The NIOSH Mining Program has seven areas of health and safety research activity, addressing respiratory diseases, hearing loss, cumulative musculoskeletal injuries, traumatic injuries, disaster prevention, rock safety engineering, and surveillance and training. NIOSH and MSHA appear to work closely together to prioritize health and safety research, with NIOSH carrying out the R&D in response to the regulatory environment established by MSHA. The 2005 NIOSH budget for mine health and safety research ($30 million) represents a decrease of approximately $12 million in nominal dollars for health and safety research compared to 1994-1995 funding.

Reclamation and Rehabilitation of Abandoned Mined Lands. The Office of Surface Mining Reclamation and Enforcement was established in the Department of the Interior in 1977 following passage of the Surface Mining Control and Reclamation Act of 1977 (SMCRA), with the primary role of regulating surface coal mining. Among the stated purposes of SMCRA were to support research, training programs, and the establishment of research and training centers in the states on various aspects of mineral production. Although the involvement of OSM with aspects of extraction research is minimal, OSM does have limited technical and applied science activities in support of its regulatory mission. In particular, OSM, in cooperation with the states, plays a major role with regard to the reclamation and rehabilitation of abandoned mined lands.

The environmental problems associated with active and abandoned mines and their abatement, particularly land reclamation and water quality maintenance, and the proper handling and disposal of the spoils and wastes from mining operations (e.g., mountain top coal mining, coal combustion residues), also receive regulatory attention from the U.S. Environmental Protection Agency. EPA is also involved in a program to promote the capture and utilization of coal bed methane. Overall, coal mining research in EPA is limited to support for its regulatory role.

Mining Regulation. The Mine Safety and Health Administration, in the Department of Labor, provides technical support and training services to its personnel and to personnel from the mining industry through its Pittsburgh Safety and Health Technology Center and the National Mine Health and Safety Academy. The direct involvement of MSHA in funding mining research is limited because of its primary regulatory role. However, MSHA undertakes field investigations, laboratory studies, and cooperative research activities on health and safety issues in support of its inspection and technical support functions. Further, MSHA evaluates new equipment and materials for use in mines at its Approval and Certification Center. It also supports state miner training activities through its state grants program.

State Government Research Programs

State government involvement in coal mining and processing research is primarily dependent on the importance of the mining industry to each particular state. The major coal-producing states—Wyoming, West Virginia, Kentucky, Pennsylvania, Texas, Virginia, and Illinois—have or have had agencies with specific responsibilities for health, safety, and environmental issues associated with coal mining. Further, mining industry organizations in these states work closely with state agencies to support research programs that address the specific needs of coal reserve estimation and coal mining operations. These state agencies also work with their corresponding federal agencies, particularly for the acquisition of federal grants to support industry's needs. Some state governments have provided grants for coal processing research in academic departments (e.g., Virginia Polytechnic Institute and State University, University of Kentucky, Southern Illinois University) or at university-affiliated research centers (e.g., University of Kentucky's Center for Applied Energy Research and West Virginia University's National Center for Coal and Energy Research).

Product Engineering by Equipment Manufacturers

The mining industry is truly international—not only are mining operations carried out globally, but there is considerable capital, knowledge, and mined-materials flow across international boundaries to satisfy the global demand for mined and processed materials. The coal industries in different countries have much in common, particularly with regard to health, safety, and environmental issues. Because of these similarities, there is considerable exchange of research results—developments in one country are quickly incorporated into mining practices in other countries. This global interaction is particularly facilitated by mining equipment manufacturers. The consolidation of coal mining equipment manufacturers over the past three decades and the broad applicability of equipment across a range of mining situations have led manufacturers to work with mining clients and their own suppliers to develop evolutionary improvements to their products. In addition, equipment manufacturers invest substantial resources to improve the durability and reliability of mining equipment. For example, one leading mining equipment manufacturer indicated that its global engineering budget was approximately $40 million, with about 25 percent spent on engineering development activities that are related mostly to evolutionary advances and software development.

Some equipment manufacturers have worked in partnership with government agencies and mining companies to develop and demonstrate new concepts (e.g., three major equipment manufacturers are members of the Australian CRCMining program; see Box 4.3). For some equipment manufacturers, mining equipment is only one of many product lines. The applied engineering research and development work that they conduct is generally fundamental to their production and materials processes, and the research is often proprietary and not generally available to the wider industry.

Other Coal Mining and Processing Research

Cross-industry research under the aegis of coal companies or coal industry organizations, or with support from industry organizations, appears to be minimal. There are no longer organizations such as Bituminous Coal Research, Inc. (BCR) that used to work on coal mining and coal preparation issues. Instead, industry's emphasis is on improvements to existing technologies—the remarkable increases of mining productivity since the mid-1970s (Figure 4.4) are a testimony to the development and adoption of evolutionary improvements in mining technology and practices. Several coal companies work in partnership with government agencies and academic institutions on coal mining research projects. The importance to researchers of access to operating mines and input from mining company experts is particularly worth noting.

BOX 4.3
Australian Government-Academia-Industry Research Models

One model for cooperative government-academia-industry research is that of the Australian Coal Association Research Program (ACARP).[1] ACARP, which began in 1992, is funded by a 5 cent per tonne tax paid by the Australian black coal industry that generates $A10 million-$A15 million annually. The funds are paid to Australian Coal Research Ltd. (ACR), a company established by the industry to manage all aspects of the program. Each year, ACARP's industry-chaired committees select approximately 80 research projects for funding from about 300 proposals submitted under a competitive solicitation. The amount of leveraged matching funding for these projects from sources outside ACARP has been at an average ratio (external:ACARP) of more than 2.5:1 over the last three years. The research projects, which are conducted by university, industrial, and government-affiliated researchers, are monitored by industry representatives.

The Australian government also supports the Cooperative Research Centre (CRC) for Mining (CRCMining),[2] an incorporated entity created in 2003. CRCMining is projected to receive $A27 million in funding from the Australian government over seven years. This will be matched by about another $A100 million of funding from 12 industry and 4 university partners. CRCMining is one of about 60 CRCs in Australia, 8 of which are concerned with mining and energy. Each CRC is incorporated and operates under a formal agreement with the Australian government of up to seven year's duration. Under this agreement, the government agrees to provide a certain level of funding each year to the CRC, and CRC participants agree to undertake certain activities and contribute specified personnel and resources.

---

[1]See http://www.acarp.com.au/index1024.shtml.
[2]See http:// www.crcmining.com.au/.

The National Mining Association (NMA) earlier outlined its vision and goals for the Mining IOF program (NMA, 1998). This recognized the importance of developing research priorities for new technologies and joint sponsorships of chosen projects, and resulted in an NMA-DOE partnership that supported several roadmaps as part of the Mining IOF program. Three specific roadmaps (*Mining Industry Roadmap for Crosscutting Technologies*, 1999; *Mineral Processing Technology Roadmap*, 2000; and *Exploration and Mining Technology Roadmap*, 2002) resulted in projects funded by DOE.

FINDINGS AND RECOMMENDATIONS— COAL MINING AND PROCESSING

The more difficult mining conditions that will be encountered in the future will require improved methods to protect the health and safety of mine workers, careful environmental management of mined lands and waste products, and improved productivity and recovery to optimize use of the nation's coal resource.

Improved Mine Worker Health and Safety

A range of factors increase health and safety risks to the coal mining work-force, including the introduction of new equipment and systems; the commencement of mining in virgin areas; the infusion of new workers; and the mining of multiple seams, seams that are thinner, thicker, or deeper than those customarily mined at present and new seams that underlie or overlie previously mined-out seams. All of these factors are likely to apply to some degree in future mines, irrespective of whether the higher production scenarios suggested in some forecasts eventuate. If they do materialize, then these risks are likely to become even more pronounced.

- There are major knowledge gaps and technology needs in the areas of survival, escape, communications systems (both surface-to-underground and underground-to-underground), and emergency preparedness and rescue. Additional risk factors that are likely to apply in the deeper mines of the future are the potential hazards related to methane control, dust control, ignition sources, fires, and explosions.
- Greater understanding and better prediction of strata behavior to prevent unanticipated[12] roof collapse, particularly problems associated with roof and side fall during thick seam extraction, are essential for maintaining and improving worker safety.
- Federal support for health and safety research significantly decreased about a decade ago, and has essentially remained constant since that time.

Recommendation: Health and safety research and development should be expanded to anticipate increased hazards in future coal mines. These R&D efforts should emphasize improved methane control, improved mine ventilation, improved roof control, reduced repetitive and traumatic injuries, reduced respiratory diseases, improved escape and rescue procedures, improved communications systems, and research to reduce explosions and fires. This should be coupled with improved training of the mining workforce in all aspects of mine safety. R&D should also be directed toward lowering the exposure of mine workers to hazardous conditions, particularly through expanded use of remote sensing and the automation of mining operations.

---

[12]Roof collapse is anticipated during longwall mining after the coal has been removed (see Appendix E).

Most mining health and safety research by the federal government is carried out by the Mining Program at the National Institute for Occupational Safety and Health. Technology-related activities in the Mine Safety and Health Administration are limited to technical support and training services for its personnel and those from the mining industry. With NIOSH carrying out the research needed to improve mine safety and to support MSHA's regulatory role, these two agencies play a vital role in coal mine health and safety. The committee estimates that the enhanced health and safety program proposed here will require an additional $35 million to provide total annual R&D funding of approximately $60 million and recommends that NIOSH continue as the lead agency with enhanced coordination with MSHA and industry.

Improved Environmental Protection

Coal mining has environmental impacts on air, water, and land. Actions taken to meet federal and state environmental regulations already respond to and ameliorate many of these effects. However, there are gaps in the knowledge base related to a range of environmental issues that need to be addressed, and future changes in environmental requirements may drive the need for new coal mining and beneficiation technology.

- As mining extracts coal from deeper and operationally more difficult seams by both surface and underground methods, it is likely that many existing environmental issues and concerns will be exacerbated. New concerns, particularly related to greater disturbance of hydrologic systems, ground subsidence, and waste management at mines and preparation plants, are likely to arise.
- Inadequate understanding of post-mining strata behavior and the associated hydrologic consequences of mining in both surface and underground mines affects mine permitting, mine development, environmental mitigation, and post-mining land use, including use for waste management.
- Research offers considerable potential to mitigate the effects of past mining practices, particularly acid mine drainage on abandoned mine lands.
- The regulatory environment (framed by such statutes as the Clean Air and Clean Water Acts) and technical support programs administered by both state and federal agencies, and implemented by mining companies through their compliance practices, are inadequately supported by existing research programs.

Recommendation: Additional research is needed to mitigate the adverse environmental impacts associated with past, existing, and future coal mining and processing. Research activities should focus particularly on developing techniques to mitigate the alteration and collapse of strata overlying mined areas, to model the hydrological impacts of coal mining, to improve mine mapping and void detection, to improve the stability of

spoils on steep slopes, and to improve the construction and monitoring of impoundments.

Both the Office of Surface Mining Reclamation and Enforcement and the Environmental Protection Agency, although primarily regulatory agencies, fund limited R&D activities in support of their missions. The committee estimates that annual funding of approximately $70 million will be required to conduct the research necessary to adequately respond to the environmental impacts of past, existing, and future mining operations. The committee recommends that OSM should be the lead agency in this effort, and it should coordinate closely with related EPA and state research activities.

Improved Mine Productivity and Resource Optimization

The productivity of U.S. coal mines increased two- to three fold in the past three decades, largely due to evolutionary improvements, most notably the introduction of longwall mining in eastern underground mines and the development of large surface operations in the West. The sustained production and productivity increases that followed these changes resulted from incremental improvements in equipment and mining practices by mining companies and equipment manufacturers, and there has been little research and development on truly advanced mining technologies.

• The development of advanced technologies, such as thin-seam underground mining technology or dry processing methods for western surface-mined coals, will present opportunities to recover a significant portion of potentially recoverable coal that currently is not extracted and may be permanently lost. In situ extraction or utilization methods, while they have not found broad application in the past, may become attractive as more easily mined reserves are exhausted. Many advanced mining technologies with the potential to reduce mine hazards, such as remote sensing, continuous monitoring at the mine face, remote control, and autonomous systems, also have the potential to increase production and productivity and improve resource recovery. Although the national coal resource is truly vast, the economically recoverable reserve base will depend on mining costs that in turn are determined by labor, environmental, and technological factors.
• Small percentage increases in coal recovery through improved coal preparation processes and improved mining methods, perhaps including in situ extraction, have the potential to significantly expand economically recoverable reserves of both eastern and western coals. The development of these technologies, increasingly needed as coal reserve quality decreases over time, will help to maximize utilization of the nation's coal resource.
• The global transfer of coal mining and processing technology within the industry is facilitated by international equipment manufacturers, who work

closely with suppliers and the larger mining clients on evolutionary product developments. However, there is little evidence of the efficient transfer of technologies from outside the mining industry. This is at least partly due to the relatively small market that the coal mining industry represents to potential technology suppliers and the scarcity of coal mining research at academic institutions and national laboratories.

• There is minimal federal support for the research and development of advanced mining technologies and practices that are necessary to optimize utilization of the nation's coal resource.

Recommendation: There should be renewed support for advanced coal mining and processing research and development to optimize use of the nation's coal resources by increasing the amount of coal that is economically minable through technological advances that accommodate health, safety, and environmental requirements. The focus of this R&D should be on increased integration of modern technology in the extraction and processing phases of coal production, with particular emphasis on emerging advances in materials, sensors, and controls; monitoring; and automated mining systems.

Although there is currently little federal funding for advanced mining technology to improve resource recovery, in the past the Department of Energy successfully partnered with the National Mining Association as part of the Mining Industry of the Future program. In addition, there are government-industry-academic cooperative models in other countries that are successful in directing and funding mining research (e.g., see Box 4.3). Research to develop advanced mining technologies requires not only cooperation among relevant federal agencies, but also participation by academic institutions as well as funding, guidance, and technology transfer by industry. The committee estimates that advanced coal mining and processing R&D will require a total of approximately $60 million per year and recommends that this funding should comprise $30 million in federal support, with cost sharing from non-federal sources. The DOE Office of Fossil Energy should be the lead federal agency and should coordinate with the National Science Foundation, OSM, NIOSH, academic institutions, and the coal industry to ensure that all research activities carefully consider the environmental, reclamation, and health and safety aspects of coal mining.

5

Transport of Coal and Coal Products

Like all fuels, coal must be transported to an end user before it can be used. Specific transportation needs vary—Gulf Coast lignite is generally transported over very short distances to minemouth power plants, Appalachian and Illinois Basin coals are typically transported over somewhat longer distances from mine to market, and coal mined in the Powder River Basin may travel distances ranging from less than 100 miles to more than 1,500 miles before it reaches the user (NCC, 2006). Therefore, growth in coal use depends on having sufficient capacity to deliver increasing amounts of coal reliably and at reasonable prices. Conversely, insufficient capacity, insufficient confidence in reliable delivery, or excessive transportation prices could reduce or eliminate growth in coal use.

With the electric power sector accounting for more than 90 percent of U.S. coal use (Table 5.1), coal transport to the more than 600 coal-burning power plant sites in the nation is especially important. Of these plants, rail transportation serves approximately 58 percent, waterborne transportation serves 17 percent, trucks serve 10 percent, 12 percent are served by multiple modes of transportation (primarily rail and barge), and 3 percent are minemouth plants with conveyor systems (NCC, 2006). In 2004, more than 85 percent of coal shipments were delivered to consumers by either rail (684 million tons), truck (129 million tons), or water (98 million tons) (EIA, 2006g; see Table 5.1). However, Energy Information Administration (EIA) statistics report only the method by which coal was delivered to its final destination and do not describe how many tons may have traveled by other means along the way—almost one-third of all coal delivered to power plants is subject to at least one transloading along the transportation chain (NCC, 2006). For example, the figure for waterborne transport does not include

TABLE 5.1 Tonnage of Coal Delivered to Consumers in 2004

Delivery Method	Electricity Generation	Coke Plants	Industrial[a] (except coke)	Residential or Commercial	Total
Great Lakes	8,644	1,144	1,341	—	11,128
Railroad	625,830	10,414	46,031	1,975	684,249
River	71,062	3,722	7,915	406	83,105
Tidewater piers	3,391	—	530	—	3,936
Tramway, conveyor, and slurry pipeline	79,997	1,014	31,975	—	115,262
Truck	73,441	453	50,266	2,741	128,900
Unknown	—	—	—	—	28,005
Total	**863,802**	**17,095**	**150,309**	**5,122**	**1,064,348**

NOTE: Figures, in thousand short tons, are for final delivery and do not reflect transloading to or from other modes during transit.

[a]This category includes coal that is transported to plants that transform it into "synthetic" coal that is then distributed to the final end user—a substantial component goes to electricity generation plants.

SOURCE: EIA (2006g).

coal that was transloaded to rail, truck, or other transport modes before final delivery, and the U.S. Army Corps of Engineers reported that 223 million tons of domestic coal and coke were carried by water at some point in the transport chain in 2004 (USACE, 2006).

Coal transportation, especially by truck and rail, affects communities through which the coal passes. Trucks hauling coal have the potential to damage roads and cause deaths or injuries in accidents. Coal trains crossing local roads temporarily block those roads, adding traffic congestion and potentially delaying or degrading responses by police, fire, and other emergency responders and temporarily cutting off some residents from emergency services (e.g., see TVA, 2005).

TRANSPORTATION BY RAIL

Coal producers and users depend heavily on rail transportation (see Figure 5.1). In 2004, rail transported 64 percent of U.S. coal shipments to their final domestic destinations and 72 percent of coal delivered to power plants (EIA, 2006g; see Table 5.1). Under the EIA's reference case forecast (EIA, 2006d), all transportation modes—particularly railroads—will be called on to transport more coal for longer distances to both existing and new markets. This forecast projects that Appalachian coal production will increase slightly (2 percent) between 2004 and 2030, production from the interior will increase by 135 million tons (92 percent), and production from the rail-dependent West will increase by 435 million tons (76 percent). Accordingly, future growth in coal use will depend on the availability of sufficient rail capacity to deliver increasing amounts of coal and on the railroad industry's ability to do so reliably and at reasonable prices.

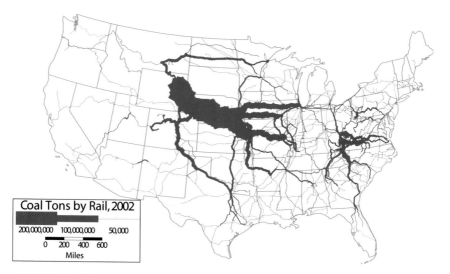

FIGURE 5.1 Schematic showing coal tonnage transported by rail in 2002 throughout the 48 conterminous states. SOURCE: Courtesy of Bruce Peterson, Center for Transportation Analysis, Oak Ridge National Laboratory.

Although continuing technology improvements may help railroads to add capacity and provide reliable delivery at a reasonable cost, it is unlikely that federally sponsored research and development will be significant contributors to such improvements—capacity, reliability, and price are all much more dependent on supply and demand, business practices, the investment climate, and regulatory oversight in the railroad industry. In addition, although the industry faces staffing constraints, worker health and safety concerns, environmental regulation, and community concerns, these issues do not threaten capacity, reliability, or price to an extent that would materially affect projections of future coal use.

Rail Capacity

Demand for transport of coal by rail has increased markedly in recent years. This is especially the case in the West, where the tonnage of coal transported on the line jointly operated by BNSF Railway Company and Union Pacific Corporation to serve the southern Powder River Basin coal fields (the "Joint Line") (Figure 5.1) increased from 19 million tons in 1985 to 325 million tons in 2005 (UPC/BNSF, 2006). Increasing rail capacity depends on capital investments for rolling stock and additional track, and such investments require confidence that demand and revenue will remain high over the long term (Hamberger, 2006). The railroads have cited changes in demographics, training requirements, and limits on the availability of qualified personnel as posing a risk to their ability to meet the long-term demand for rail service (BNSF, 2005; UPC, 2006).

Rail Reliability

Weather and other natural phenomena, such as earthquakes, fires, and floods, have the potential to cause localized line outages that can, in turn, adversely affect an entire rail network. Weather conditions in Wyoming in May 2005 demonstrated this risk when heavy rain and snow, combined with accumulated coal dust in the roadbed, led to track instability on the Joint Line (UPC, 2006). Two coal trains derailed on consecutive days, damaging the line and temporarily putting it out of service (EIA, 2005b). Both Union Pacific and BNSF declared *force majeure*, beginning with the derailments and continuing until normal operations were restored. Track maintenance and restoration disrupted operations and reduced shipments on the Joint Line throughout most of the rest of 2005 (UPC, 2006). The spot price of Powder River Basin 8,800 Btu (British thermal unit) coal reflected the severity of this disruption, rising from $8.19 per short ton just before the derailments to $16.89 per short ton in October 2005 (EIA, 2005a, 2005c).

The terrorist attacks of September 11, 2001, and the more recent attacks on passenger transportation systems in London, Madrid, and Mumbai, have raised concerns about possible terrorist disruptions of freight rail transportation. Even when freight rail infrastructure is not directly the target of a terrorist attack, government efforts to protect against such attacks can slow trains, increase congestion, and adversely affect railroads' profitability, financial condition, or liquidity (UPC, 2006).

State utility regulators have noted increases in uncertainty associated with the availability of rail cars for loading the coal at its point of origin, the availability of locomotive power, and the arrival time at the train destination (NARUC, 2006). Opinions differ about whether or not disruptions in coal delivery reflect a substantial and ongoing problem and about whether the power plant operators or the railroads should modify their activities to respond to these delivery problems (English, 2006; Hamberger, 2006; McLennan, 2006; Mohl, 2006; Wilkes, 2006).

The rail networks that transport the nation's coal—like air traffic control and electric transmission networks—have an inherent fragility and instability common to complex networks. Because concerns about sabotage and terrorism were largely ignored until recently, existing networks were created with potential choke points (see Figure 5.1) that cause vulnerability. The complex and dynamic interactions between societal and environmental factors—as well as the intrinsic dynamics of a system that operates close to its capacity—result in the potential for small-scale issues to become large-scale disruptions.

Price

The Staggers Rail Act of 1980 removed many regulatory restraints on the railroad industry. The Staggers Act allowed the Interstate Commerce Commission (now succeeded by the Surface Transportation Board) to regulate rates only when competition is not sufficient to keep rates below a statutory threshold expressed as a multiple of the railroad's variable cost (FRA, 2004). Since the Staggers Act took effect, a long-term decline in railroad market share has been reversed and freight rates (adjusted for inflation) have declined by 1 to 2 percent annually (FRA, 2006).

However, developments since 1980 have significantly reduced competition in the industry. More than 40 Class I railroads (a railroad with at least $250 million in operating revenues in 1991 dollars) served North America in 1980, and only 7 remain today. Of these, two railroads in the West (Union Pacific and BNSF) and two in the East (CSX and Norfolk Southern) control more than 95 percent of the rail business. Consequently, each of the coal supply regions—the Powder River Basin, Illinois Basin, and Appalachian regions—is served by only two railroad companies for coal transport to power plants (NARUC, 2006).

The combination of reduced rail competition, perceived problems in the delivery of coal by rail, and price increases associated with the 2005 rail disruptions has caused some concern on the part of coal-fired power plant owners about both the reliability and the price of coal delivery. Severe and frequent delivery problems or spikes in prices have the potential to reduce future coal use by affecting the climate for coal-fired power plant investments.

TRANSPORTATION BY TRUCK

More than 12 percent of the total coal transported in 2004 in the United States—about 129 million tons—was moved by truck (EIA, 2006g). Typical truck haul lengths (one way) are less than 100 miles, averaging about 32 miles. Significant tonnages of coal are trucked in some states, most notably West Virginia (18 million tons shipped annually) and Kentucky (17 million tons trucked annually). Truck shipments are also an important component of multimodal coal transport in Kentucky. The issues associated with truck transport are primarily associated with road maintenance, the generation of noise and dust, and traffic safety.

WATERBORNE COAL TRANSPORTATION

Transportation on the inland waterways and Great Lakes is an important element of the domestic coal distribution system, carrying approximately 20 percent of U.S. coal tonnage and making 10 percent of deliveries to end-use consumers. The amount of waterborne transported coal, approximately 306 million tons in 2004 (including imports and exports), has remained relatively constant over the last two decades. Coal represents a significant share of shipping on the inland waterways, accounting for approximately 20 percent of total cargo. Barge transportation rates on contract coal shipments are about one-half to two-thirds those of rail haulage on a ton-mile basis, and truck transportation rates are an order of magnitude higher than waterborne transportation rates (EIA, 2006h).

Barge traffic is particularly important in the midwestern and eastern states, with 80 percent of shipments originating in states along the Ohio River. This reflects the large number of coal mines and electricity generation facilities that have barge loading and unloading facilities along the Ohio River and its tributaries. Some coal exports from the United States to Canada also move across the Great Lakes. These exports have decreased in recent years, but lake traffic has remained approximately constant because of increased movement of Powder River Basin (PRB) coal shipped between U.S. ports. Like PRB coal, which is transloaded from rail to lake vessel or barge, much waterborne coal is transloaded before final delivery to the ultimate consumer. Although total domestic waterborne coal cargo is about 200 million tons, only about half of that coal (110 million tons) is finally delivered by water to its final customer (Table 5.1), principally to electricity generating facilities.

Maintenance of the critical infrastructure along the inland waterways and Great Lakes (i.e., locks and dams, dredging of ports) is the responsibility of the U.S. Army Corp of Engineers (USACE). USACE construction and rehabilitation projects are funded on a 50-50 cost-shared basis from appropriations and from the Inland Waterways Trust Fund, established in 1986, which derives its revenue from a 20-cent-per-gallon tax on fuel used for commercial waterway transportation. Between 1992 and 2001, congressional appropriations were less than Inland Waterways Trust Fund income and therefore the fund balance grew, a situation that began to be reversed in 2005 with greater administration requests and congressional appropriations. The USACE also spends about $500 million per year on operation and maintenance (O&M) of the waterway system, of which $135 million is spent in the Ohio River and Great Lakes Division.[1] O&M expenditures for the total system have been essentially level (in constant dollars) since the 1970s, below levels that the industry believes are optimum for the aging system.

The use of Inland Waterways Trust Fund money has been a source of considerable concern within the barge and towing industry (Knoy, 2006). Similarly, the operators of commercial shipping on the Great Lakes have warned that inadequate port dredging is hampering the transport of coal from the Powder River Basin (LCA, 2006). The USACE and Congress receive recommendations for the use of the trust fund from the Inland Waterways Users Board (IWUB), an 11-member industry advisory committee, and this body recently warned that

---

[1]Presentation to the committee by John Moran, Waterways Council, Inc., June 2006.

deferred maintenance has resulted in serious structural failures (IWUB, 2005). The IWUB identified approximately 20 construction and rehabilitation projects that it believes are critical to the inland waterways systems, with a total price tag to completion of more than $5.4 billion. This would require an annual expenditure of $477 million, about $100 million above the FY 2006 actual appropriation. However, this level of expenditure would not be sustainable with the current trust fund balance and expected future income (IWUB, 2005). Fund income has averaged about $100 million per year over its history, and it would be required to contribute $2.7 billion as its share of the $5.4 billion called for by the IWUB.

TRANSPORTATION OF COAL EXPORTS AND IMPORTS

Exports of coal from the United States are currently around 50 million tons, a little less than half of the record export tonnages transported in the 1980s. Exports are expected to decrease in the future, primarily due to the anticipated availability of low-cost coal supplies from South America, Asia, and Australia (EIA, 2006d). In fact, the EIA reference scenario predicts that the U.S. share of the total world coal trade will fall from 6 percent in 2003 to 3 percent in 2025. At the same time, U.S. imports of low-sulfur coal are projected to grow, from the current 28 million tons to almost 90 million tons by the year 2030. The potential need to meet tighter emissions targets may make coal imports an attractive option for coal-fired power plants in the Gulf Coast and Atlantic seaboard areas (EIA, 2006d).

The national transportation network is not expected to be challenged by these predicted export and import trends. Transloading terminals on the Gulf Coast and the Atlantic seaboard have adequate capabilities for managing such traffic, and they have managed increased volumes in the past. However, reversing or shifting the flow direction from export to import may present logistical and operational problems for the transportation infrastructure, principally the railroads.

ELECTRICITY TRANSMISSION

Constraints on the delivery of electricity from power plants can reduce the natural competitive advantage that coal-fired power plants have over plants fueled by oil or natural gas that cannot generate electricity as cheaply. Consequently, transmission constraints have the potential to limit future coal use. Coal's competitive advantage relies on "economic dispatch"[2] that theoretically operates in the electricity generation market (DOE, 2005). However, in practice

---

[2]Every power plant has a schedule of production levels and costs. In theory, units are called upon to provide power in "merit order," in which the least expensive units are dispatched first, with additional units being dispatched in order of increasing costs until electricity needs are met. Factors that could increase the production costs of coal-based plants and thereby alter dispatch order, such as environmental constraints, are discussed in Chapter 6.

it is frequently necessary to dispatch units out of merit order because of electric transmission infrastructure limitations, and in some cases this results in higher-cost resources being dispatched in place of lower-cost resources (DOE, 2005). When this happens, some lower-cost generators lose opportunities for sales.

Much of the nation's coal-fired electric generating capacity is located at some distance from the urbanized areas that have the largest and most concentrated demands for electricity. For example, PJM Interconnection LLC operates the world's largest centrally dispatched transmission grid, stretching from Illinois to New Jersey and extending as far south as Virginia and North Carolina (PJM, 2007). The bulk of the lower-priced coal-fired generation for the PJM grid is located in northern West Virginia, northern Virginia, Maryland, eastern Ohio, and southwestern Pennsylvania. The eastern Mid-Atlantic portion of PJM's territory, which includes New Jersey, Delaware, southeastern Pennsylvania, and eastern Maryland, has experienced growing customer demand and relatively little new generation capacity. To the extent that transmission capability allows, lower-priced coal-fired generation in the central part of PJM's territory displaces higher-cost generation in the East. However, both technical (e.g., risk of over-heating transmission lines) and operational (e.g., need to maintain voltage at minimum levels) factors limit transmission capability.

Significant portions of the country are subject to transmission congestion and the resulting out-of-merit dispatch. Two densely populated and economically vital areas—the Atlantic coastal area from metropolitan New York to northern Virginia, and Southern California—currently have major transmission congestion or are projected to suffer severe congestion effects in the future (DOE, 2006, 2007b). The severity of such effects is linked to the size of the population affected, economic costs, size of the reliability problem, impact of a grid failure on the nation, or some combination of these factors. Four areas of concern were identified in which a large-scale congestion problem exists or may be emerging—New England, the Phoenix-Tucson area, the San Francisco Bay area, and the Seattle-Portland area. This analysis also noted the likely need for significant additional transmission investments to enable increased flows of electricity from midwestern coal-fired plants into the PJM grid and New York (DOE, 2006).

Planning for reliable electricity in the areas of greatest demand depends on a combination of local power plants to meet local demand without undue stress on the transmission system; distributed resources such as small on-site generators, energy efficiency and other demand reduction; and new or upgraded transmission infrastructure (NYC, 2004). It is difficult to predict the extent to which particular urbanized regions will endeavor to enhance the reliability of their electricity supply through local generation and transmission or by instituting energy efficiency or other demand reduction measures. If these areas implement alternative ways to increase electricity supply or enhance supply reliability other than by relying on new and upgraded transmission infrastructure, the need for increased coal usage will be diminished.

As is the case with rail transportation, the electric transmission system can be vulnerable to initially localized disruptions that ultimately have severe and widespread impacts. For example, the failure to manage tree growth along transmission rights-of-way was cited as the root cause of an August 2003 blackout that affected Ohio, Michigan, Pennsylvania, New York, Vermont, Massachusetts, Connecticut, and Ontario, with estimated costs ranging from $4 billion to $10 billion in the United States alone (DOE/NRC, 2004).

TRANSPORT OF COAL-DERIVED PRODUCTS

In the future, transport of a range of coal-derived products also may require attention. For example, liquid fuels and substitute natural gas derived from coal are being assessed with increased interest as a result of recent oil and gas price increases and national security concerns. Some of the coal use scenarios described in Chapter 2 include projections for growth in coal-to-liquids and coal-to-gas plants in the post-2020 period. In general, the transport of energy products from such plants would be similar to the pipeline and other distribution systems currently employed at petroleum refineries or gas processing plants. However, should a significant coal-based synthetic fuels industry begin to materialize in future decades, issues related to the transport of energy products from such facilities may require further research.

If geological sequestration of CO_2 is implemented on a large scale as a greenhouse gas mitigation measure in the future, it will be necessary to transport large quantities of CO_2 from their sources to geological storage sites. Ideally, CO_2 sequestration would take place at sites in close proximity to the sources of CO_2, generally coal-based power plants or other large industrial facilities that capture and compress CO_2 for transport and storage. However, not all coal plants are located immediately above or adjacent to geologic storage sites. In such cases, transport of the CO_2 by pipeline would likely be the most economical and preferred method, although it is also possible to transport CO_2 in road tankers, rail tankers, or ships (in cases where the sequestration site is located far offshore) (IPCC, 2005). The proximity of potential sequestration reservoirs will need to be considered, along with many other factors (e.g., proximity to coal fields, transport costs, electricity delivery costs, availability of water), when sites for power plants are evaluated. An extensive description and analysis of CO_2 transport is presented by the Intergovernmental Panel on Climate Change (IPCC, 2005), and the following two paragraphs are derived from that report.

Currently, there are more than 2,500 km (~1,500 miles) of long-distance CO_2 pipelines operating in the western and southern United States. These pipelines transport more than 40 megatons of CO_2 per year,[3] primarily from natural CO_2

---

[3]This amount compares to approximately 2,000 megatons of CO_2 emitted from all U.S. coal-fueled power plants in 2005.

wells and from one coal gasification facility, for use in enhanced oil recovery (EOR). The oldest of these pipelines was completed in 1972, and the longest is 808 km (Gale and Davidson, 2004). In the 13 years from 1990 to 2002, there were 10 incidents of leakage involving CO_2 pipelines, with no injuries or fatalities (Gale and Davidson, 2004). Existing CO_2 pipelines are located in rural areas of low population density and, unlike natural gas pipelines, do not pose a risk of combustion or explosion.

The composition of CO_2 pipelines and their manufacture, construction, maintenance, and operation are all mature technologies. Most existing pipelines carry reasonably pure CO_2, although some also contain impurities (e.g., H_2S derived from petroleum refining). Depending on the CO_2 source and capture technology, some future sequestration pipelines might contain various amounts of other impurities such as SO_2, NO_x, oxygen, and nitrogen, possibly requiring some modification to current pipeline design specifications. It is expected, however, that allowable levels of impurities will be determined by future regulatory requirements governing CO_2 sequestration.

FINDINGS—TRANSPORT OF COAL AND COAL PRODUCTS

The issues associated with the transport of coal and coal-derived products are related primarily to the regulatory and business environments, and with the exception of an improved understanding of complex networks, there seems to be little requirement for research activity. Accordingly, the committee finds the following:

• The greater coal use projected in some of the scenarios discussed in Chapter 2 will be possible only if sufficient transport capacity is available to reliably deliver the increased amounts of coal at reasonable prices.

• Transport of coal by rail and by waterway will be critical for increased coal use. The capacity, reliability, and price of rail transportation—the dominant mode of coal transport—depend largely on the supply and demand for rail transportation, as well as on prevailing business practices, the investment climate, and the nature of regulatory oversight of the railroad industry. The capacity, reliability, and price of rail transportation of coal depend to a far lesser degree on research and development. Reliable and sufficient waterborne transportation—the second most prevalent method of coal transport—depends on the construction and maintenance of waterway infrastructures, especially lock-and-dam infrastructure and port capacity.

• Much of the nation's coal-fired electric generating capacity is located at some distance from the urbanized areas that have the largest and most concentrated demands for electricity. Projections of higher coal use depend on sufficient capacity to transmit electricity from coal-based power plants to such areas reliably and at a reasonable cost. Conversely, the projected increases in coal use will diminish if those high-demand areas satisfy much of their growing demand for electricity not by expanding their ability to import electricity from areas where coal is plentiful, but by a combination of energy efficiency, demand response, and local electric generation from sources other than coal.

Both the rail transportation and the electric transmission systems are complex networks in which localized disruptions can have severe and widespread impacts. Research is needed to better understand the factors that control these large and complex networks to minimize the risks of cascading system disruptions.

6

Coal Utilization

This chapter addresses key issues associated with the utilization of coal in the United States. As noted in Chapter 1, although the committee's task is broad and encompasses the entire coal fuel cycle, the congressional framers of this study requested that it focus primarily on the "upstream" aspects of the coal fuel cycle, recognizing that the research and development (R&D) aspects of coal utilization technologies have been extensively studied and assessed in prior reviews by the National Research Council and others. Accordingly, this chapter is not intended to provide a comprehensive review of R&D activities related to coal utilization. Rather, it presents a brief overview of coal use technologies and related issues and then concentrates on those factors associated with coal use that are likely to impose constraints on future demands for coal.

Of the many factors that influence the nation's use of coal as an energy source, the analysis presented in Chapter 2 indicates overwhelmingly that the environmental impacts of coal use—especially carbon dioxide emissions associated with global climate change—pose the greatest potential constraint on future coal utilization. Furthermore, the uncertainty about future requirements to control these environmental impacts can itself act as a constraint on future coal utilization. Therefore, this chapter focuses primarily on these environmentally related issues, noting current R&D programs and describing priority R&D needs to minimize these impacts.

COAL UTILIZATION TECHNOLOGIES

Figure 1.7 portrayed the sustained growth in coal use for electric power generation over the past half-century, together with the decline in use of coal as an energy source for industrial processes, transportation, and residential or commercial buildings. In 2005, the dominant use of coal in the United States (92 percent of the total) was for electric power generation (Table 6.1). Coke plants (2 percent) and other industrial uses (5 percent) account for most of the remainder, with a small amount of coal still used in residential and commercial buildings (EIA, 2006c).

Extensive discussions of the technologies used in each of these sectors, as well as the status and needs of ongoing R&D, can be found in other reports (e.g., NRC, 1994, 1995, 2000, 2002a, 2002b). Accordingly, this chapter presents only a brief overview of the major coal utilization technologies that influence the projections for future coal use in Chapter 2.

Electric Power Sector Technologies

About 90 percent of the 313,000 megawatts (MW) of coal-fired generating capacity in the United States today is based on combustion of pulverized coal (PC). This process involves reducing the coal to a powder that is burned in a boiler to generate high-pressure, superheated steam that drives a turbine connected to an electric generator (Figure 6.1). The steam is then condensed back to a liquid and returned to the boiler to repeat the cycle. Although this process can use a variety of coal types, boilers capable of burning a more uniform quality of coal are generally less expensive than those designed for a broader range of coals. Consequently, it is common for many run-of-mine coals to be cleaned to reduce coal ash and (to a lesser extent) sulfur content, thereby providing a more uniform fuel supply with a higher heating value (see Chapter 4). Several of the newer coal-fired units constructed in recent years have employed atmospheric fluidized bed combustion (AFBC) technology offering greater flexibility in fuel quality.

The overall efficiency of PC power generation is affected by many factors, including the thermodynamic cycle design, steam temperature and pressure, coal particle size (coal grind), combustion air-to-fuel ratio, fuel mixing, air leakage into the system, cooling (condenser) water temperature, and the "parasitic" energy loads required to power auxiliary equipment such as grinding mills, pumps, fans, and environmental control systems. Since 1960, the average thermal

TABLE 6.1 Coal Use in the United States by End-Use Sector in 2005

End-Use Sector	Coal Use (thousand short tons)
Electric power sector	1,037,485
Coke plants	23,434
Other industrial processes	60,340
Residential and commercial	4,217
Total	**1,125,476**

SOURCE: EIA (2006c).

efficiency of PC plants in the United States has remained in the range of 33 to 34 percent (higher heating value [HHV]), with the newest plants having efficiencies between 35 and 37 percent. The most efficient coal plants are supercritical steam units that operate at higher temperature and pressure conditions than current subcritical steam units and achieve net efficiencies in the range of 40 to 45 percent (IEAGHG, 2006). Although few supercritical units have been built in the United States, many of the 154 new PC projects proposed as of 2006 are expected to be supercritical units (NETL, 2007).

A number of proposed projects use technologies based on coal gasification rather than coal combustion. Integrated gasification combined cycle (IGCC) systems generate electricity by first converting coal or other feedstock into a clean gaseous fuel that is then combusted in a gas turbine to generate electricity (Figure 6.1). The exhaust from the hot gas turbine is used to generate steam that drives a second turbine to generate additional electricity. Like PC plants, the overall thermal efficiency of an IGCC system depends on many factors, including the

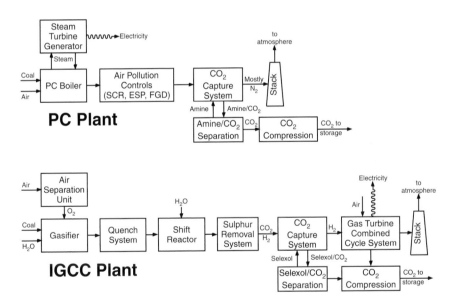

FIGURE 6.1 Schematic showing power plant designs, based on current technologies, for (*top*) a pulverized coal (PC) combustion plant with post-combustion CO_2 capture, and (*bottom*) an integrated gasification combined cycle (IGCC) plant with pre-combustion CO_2 capture. The PC plant also has emission controls for NO_x (SCR = selective catalytic reduction system), particulate matter (ESP=electrostatic precipitator), and SO_2 (FGD = flue gas desulfurization system); the IGCC plant has emission controls for NO_x, particulates and SO_2 (via quench and sulfur removal systems). SOURCE: Rubin et al. (2007).

gasifier type, coal type, oxidant type, and level of plant integration. The range of efficiencies for current IGCC systems is comparable to that for current PC plants, with significant future improvements expected as gas turbine technology develops (NRC, 2000). Another advantage cited for the IGCC process is its ability to achieve lower levels of air pollutant emissions than PC systems, although modern technology permits emissions from both types of plants to be controlled to very low levels for all regulated air pollutants (EPRI, 2006a). Both PC and IGCC plants also can achieve high levels of CO_2 removal, should that be required in the future (see below).

The gasification process is an established commercial technology that is widely used in the petroleum and petrochemical industries to convert carbon-containing feedstocks (such as coal, petroleum coke, residual oil, and biomass) to a mixture of carbon monoxide and hydrogen, referred to as synthesis gas (or syngas). For IGCC applications, the syngas is burned in a combined cycle power plant to generate electricity. More than 160 gasification plants are currently in operation worldwide, producing syngas sufficient to power the equivalent of about 20,000 MW of electrical power (IPCC, 2005). Since 1995, four coal-fueled IGCC power plants, each with an output of approximately 250 MW, have been built as part of government-sponsored demonstration programs—two in Europe and two in the United States. Although the cost of these plants was significantly higher than that of an equivalent PC plant, the experience gained from their operation, as well as from the design, construction, and operation of other industrial IGCC plants, is expected to significantly reduce future costs.

A key attraction of IGCC technology is that the incremental cost of capturing CO_2 emissions (in addition to regulated air pollutants) is lower than that for a comparably sized PC plant based on current commercial CO_2 capture technology. Thus, while an IGCC plant without CO_2 capture is currently more costly than a PC plant, an IGCC plant is typically less costly if CO_2 capture is added to coal-based power plants using bituminous coal (IPCC, 2005). However, many other factors—especially coal type—also affect the relative costs of IGCC and PC power plants. For low-rank coals (subbituminous coal and lignite), preliminary studies indicate that the current cost of an IGCC plant with CO_2 capture increases to levels comparable to or higher than the cost of a PC plant with CO_2 capture using the same coals (IPCC, 2005; EPRI, 2006b; Rubin, 2007). While there is still uncertainty about the actual cost of power plants with CO_2 capture, the potential for future constraints on greenhouse gas emissions is an increasingly important consideration when comparing the merits of investments in different power generation technologies. Although the properties, availability, and cost of different coal types is an important factor, the potential long-term impact of CO_2 capture and storage on the demand for different types of coal (in conjunction with other environmental control requirements) is unknown (and unknowable) at this time.

Industrial Sector Technologies

Coal utilization technologies in the U.S. industrial sector today fall into two main categories—coke plants that produce high-purity carbon for use in steel making, and combustion technologies that use coal to provide heat and power for industrial process operations (Table 6.1). In addition, a small amount of coal is currently used to produce substitute natural gas and high-value products such as chemicals and liquid fuels. These latter applications are projected to grow significantly over the next several decades in some of the scenarios discussed in Chapter 2.

The production of coke from coal is a centuries-old technology in which low-ash "metallurgical" coal is heated in an oven to drive off volatile matter, leaving a high-purity carbon product that is used in blast furnaces to produce iron for steel making. Modern coke plants consist of a battery of long, narrow, brick-lined rectangular ovens into which coal is fed. The volatile gases driven off by heating are collected, cleaned, and used as fuel. The hot coke product is pushed out of the oven into a rail car, quenched with water to cool it, and then shipped for use in steel making.

Most of the coal currently used at industrial facilities is burned in boilers to generate steam, just as is done in power plants. In many cases, electricity also is generated for on-site use, often in combined heat and power (CHP) systems (known also as co-generation) that yield high overall efficiencies, on the order of 80 percent or more. Industrial boilers are generally smaller than modern power plant boilers but use many of the same environmental control technologies found at larger facilities.

The recent rise in world oil prices, as well as domestic natural gas prices, has stimulated renewed interest in the production of gaseous and liquid fuels from coal. Coal liquefaction technology has long been used to produce high-quality transportation fuels, most notably in South Africa, which boasts the largest commercial facility in the world (the Sasol Group). Substitute natural gas (SNG[1]) also can be produced from coal, and one commercial plant has been operating in the United States since the 1980s (see Box 6.1). In both cases, coal gasification is a key technology. By adjusting the ratio of carbon monoxide and hydrogen in the syngas product, either gaseous or liquid products can be manufactured with the proper choice of catalysts and operating conditions (NRC, 1995, 2000, 2003a). However, the overall thermal efficiency of these processes is relatively low, on the order of 50 percent. Thus, total CO_2 emissions for a coal-to-liquids plant, including CO_2 from the conversion process and from combustion of the liquid fuel, are roughly twice that of diesel fuel produced from petroleum.

---

[1]Also known as synthetic or synthesis natural gas.

BOX 6.1
The Great Plains Synfuels Plant

The Great Plains Synfuels Plant near Beulah, North Dakota, operated by the Dakota Gasification Company, may provide a hint of an important future use for coal in the United States—the production of synthetic liquid and gaseous fuels. The plant was built with government support over two decades ago during a time of high oil prices and natural gas shortages. Each day, it uses 18,000 tons of lignite to produce about 160 million standard cubic feet of methane gas. It also produces about the same amount of CO_2, which is captured and piped 200 miles north to Saskatchewan, Canada, where it is used for enhanced oil recovery (EOR) in the Weyburn Field. An additional, roughly comparable amount of CO_2 is discharged to the atmosphere. The plant also produces anhydrous ammonia and ammonium sulfate for agricultural use, as well as a variety of other minor products. With the high price of natural gas over the past several years, the plant has been profitable and is paying off its debt to the federal government.

The Great Plains Synfuels Plant is a minemouth plant with an associated power plant. Water for the plant is pumped from the Missouri River, 20 miles to the north. The plant employs oxygen and steam-blown Lurgi gasifiers of the type used by Sasol in South Africa. Compared to natural gas from a conventional gas well, the Great Plains plant emits about twice as much CO_2 to the atmosphere per British thermal unit (Btu) of methane produced, even after subtracting the CO_2 captured and used for EOR (but including the carbon contained in the methane product). The CO_2 emitted from a coal conversion plant can be reduced to a negligible quantity if hydrogen and electricity are the co-products (rather than methane or liquid fuels) and if all of the carbon dioxide is captured and sequestered (as is planned by the Department of Energy for the FutureGen plant; see p. 101-102).

A number of studies have analyzed longer-term scenarios—to 2050 and beyond—in which coal to liquids, coal to substitute natural gas, and coal to hydrogen might begin to play increasingly important roles in the U.S. and global energy economies (e.g., IPCC, 2001; NRC, 2004a). There remains significant uncertainty, however, about these (or other) longer-term scenarios. In the time frame specified for the current study, to 2030, the scenarios reviewed in Chapter 2 suggest that power generation will remain the dominant use of coal. The remainder of this chapter focuses on issues associated with this use of coal.

ENVIRONMENTAL IMPACTS OF COAL USE

Coal-based power plants emit air pollutants and create solid and liquid wastes that can adversely affect air quality, terrestrial and aquatic environments, water resources, human health, and climate. Environmental management technologies that are already widely available and in use can reduce most of these impacts. However, no such measures are yet commercially deployed or have been demonstrated that can substantially reduce emissions of carbon dioxide from large coal-fired power plants, to ameliorate their contribution to global warming. Accordingly, the greatest difficulty in projecting future coal usage arises from uncertainty about the nature of future government action to limit CO_2 emissions from power plants and the viability of large-scale applications of technology to capture and sequester CO_2 emissions from power plants and other coal-based energy conversion facilities. Consequently, any projections of a substantial future expansion of coal use in the context of a meaningful CO_2 reduction mandate depend on the development and widespread deployment of technology to reduce CO_2 emissions from coal-based power plants.

Impacts on Air Quality

Atmospheric emissions of SO_2, NO_x, and particulate matter (PM) from coal combustion are significant sources of air pollution that have been linked to respiratory and other human health problems (e.g., NRC, 2005a; EPA, 2006a). However, as a result of a series of actions following the 1990 Clean Air Act Amendments, emissions of NO_x and SO_2 from coal-fired power plants in the United States are being substantially reduced even as U.S. coal consumption is increasing. As other requirements of the 1990 amendments are implemented, emissions of hazardous air pollutants such as mercury (EPA, 1997, 1998; NRC, 2002c) from coal-fired power plants are also expected to decline.

Projections of future coal use already reflect current government regulations aimed at reducing emissions of NO_x, SO_2, particulate matter, and mercury from coal-fired power plants. Both the technology needed to satisfy the regulatory requirements for NO_x and SO_2 emissions and the technology to eliminate more than 99 percent of PM emissions are currently in widespread commercial use (NESCAUM, 2005). The technology to significantly reduce mercury emissions in compliance with federal regulations is also expected to be commercially available by 2010 (STAPPA/ALAPCO, 2005; Feeley, 2006).

The present availability and continuing evolution of technology to respond to emissions of these pollutants make it unlikely within the 25-year time horizon of this study that current or future governmental requirements to control such emissions will materially affect future coal usage overall, although air quality considerations will continue to affect the siting of new coal-based facilities. Requirements to control air pollution from power plants are generally structured in one of two ways—as specific standards imposed on each individual electric generating unit in a regulated class; or as a "cap-and-trade" program (Box 6.2), which limits the aggregate emissions from all units in the regulated class but does not impose limits on individual units. Each approach can shift demand—either intentionally or unintentionally—to coals with differing characteristics (e.g., from bituminous to subbituminous coal or vice versa).

BOX 6.2
Cap-and-Trade Regulation of Coal-Fired
Power Plant Emissions

A cap-and-trade program requires each regulated unit to hold and use an emission credit, called an "allowance," for each ton of the regulated pollutant that the unit emits. The total number of allowances created for each year (or other compliance period) equals the total number of tons of emissions allowed under the cap, ensuring that aggregate emissions cannot legally exceed the cap. Ideally, a cap-and-trade program leads each unit to choose the most cost-effective compliance strategy. For example, to comply with federal cap-and-trade programs to reduce power plant emissions of SO_2, coal-fired power plants have a variety of strategies available:

• They can buy more allowances from sources willing to sell excess allowances in the market.
• They can install one of two main types of flue gas desulfurization (FGD) technology—either a "wet scrubber," which can reduce SO_2 emissions by more than 98 percent, or a "dry scrubber," which can reduce those emissions by 90 percent or more.
• They can switch to a fuel with lower sulfur content, which reduces SO_2 emissions in proportion to the reduction in sulfur content.
• They can combine two or more of these strategies.

The 2005 Clean Air Interstate Rule (CAIR) stipulates that beginning in 2009, each unit in the affected region of the country must hold two allowances for each ton of SO_2 it emits (EPA, 2005), rather than the one allowance per ton currently required. This effectively halves the emissions cap for the CAIR region.

In response to the 1990 Clean Air Act Amendments' requirements to reduce sulfur dioxide emissions, most existing plants chose not to install scrubbers and instead relied on a less costly compliance approach that depended largely on a switch to less expensive, lower-sulfur fuel. This choice has been cited as a key reason why the rate of growth in the use of low-sulfur coal from the Powder River Basin has far outpaced the rate of growth for higher-sulfur coal from the eastern part of the United States. In the future, in order to comply with more stringent air quality requirements for fine particulate matter (which is formed when SO_2 or NO_x react with other substances in the atmosphere), the installation of flue gas desulfurization (FGD) systems at coal-fired power plants is projected to roughly double from its current level of about 30 percent of coal-fired capacity (EPA, 2005). Whether this will significantly affect the use of western low-sulfur coals as an air quality compliance strategy remains to be seen.

Impacts on Terrestrial and Aquatic Environments and Water Resources

Coal combustion and gasification processes generate ash and other solid residues, as well as liquid effluent. Solid residues retain trace elements that were originally present in the coal, and some of these elements (e.g., arsenic, cadmium, cobalt, copper, iron, lead, mercury, molybdenum, nickel, scandium, selenium, uranium, zinc) have the potential to impact human health and the environment. Leaching of ash and slag, and the water used for handling ashes as slurries, can create water pollution problems. The processes applied to control air pollution from coal combustion also create potentially harmful residues (e.g., the removal of SO_2 from flue gases generates wastes containing sulfur compounds). Appropriate management of these residues can minimize their potential for negative effects (e.g., sludge can be dewatered by evaporation in lined ponds to reduce the potential for leakage into the soil). In addition, about 40 percent of coal combustion products produced annually are currently used for beneficial purposes, and this figure is expected to rise as a result of government and industry awareness initiatives (ACAA, 2005).

The use of water in "once-through" cooling systems in power plants can also harm aquatic life. When water is withdrawn from rivers and other water bodies for use in condensers, aquatic life can be harmed by impact against the cooling water intake structures, by being drawn into the condenser and subjected to high heat, or by higher temperatures in the water body where the cooling water is discharged.

Increases in coal combustion and gasification are likely to lead to an increase in solid residues, liquid effluents, and the use of water for condenser cooling and other plant functions. However, technologies and practices to manage the associated environmental issues currently exist and are continually being improved. Consequently, these issues are unlikely to pose a significant obstacle to increased coal use, although site-specific factors could be important in the viability and design of coal-based energy plants.

Impacts on Global Climate

Burning coal and other fossil fuels for energy releases carbon dioxide to the atmosphere, and atmospheric CO_2 concentrations have increased by about one-third since 1750; substantially higher levels are projected for the coming decades (NAS, 2005). These increased levels have been linked to global climate change and an overall warming of the earth's surface (NRC, 2001; IPCC, 2007).

In the United States, power plants account for about 40 percent of the nation's CO_2 emissions (EPA, 2006b). Strategies available to reduce these emissions include the following:

- Reducing the demand for electricity, for example, by improving the efficiency with which machinery and appliances use electricity.

• Generating electricity in plants that are more efficient. For example, a new supercritical pulverized coal plant requires less than 9,000 Btu (British thermal units) of coal energy input to generate 1 kWh of electricity (>38 percent efficiency) compared to the current average of more than 10,000 Btu/kWh of electricity (<34 percent efficiency).

• Switching to a less carbon-intensive fossil fuel. For example, generating 1 million Btu of energy as heat from coal combustion releases about 200 pounds of CO_2, whereas generating the same amount of energy as heat from natural gas releases about 120 pounds of CO_2.

• Combining fuel switching and more efficient generating plants. For example, a new natural gas-fired combined cycle (NGCC) power plant requires only about 6,800 Btu of energy input to generate 1 kWh of electricity (~50 percent efficiency). Hence, to generate 100 kWh of electricity, a NGCC plant would emit about 80 pounds of CO_2 compared to the average coal plant that emits 200 pounds per 100 kWh, or a more efficient PC plant that emits 180 pounds per 100 kWh.

• Generating electricity without using fossil fuels, for example, by using renewable resources or nuclear power.

• Capturing and sequestering (storing) CO_2 produced by fossil fuel combustion or gasification.

The technologies currently offered commercially to capture power plant CO_2 emissions can achieve net emission reductions of 85 to 90 percent at new PC or IGCC power plants (IPCC, 2005). Such technologies are widely used in a variety of industrial processes, mainly in the petroleum and petrochemical industries, but are not yet deployed commercially in the electric power sector. Although post-combustion CO_2 capture systems employing amine sorbents have demonstrated effective removal of CO_2 from flue gas streams from gas-fired and coal-fired boilers, they have not yet been applied at the larger scales typical of modern power plants (e.g., plants that generate several hundred megawatts of electricity). The same is true for the "pre-combustion" capture technologies used commercially at gasification-based processes. Various types of oxyfuel combustion systems[2] also are being developed to facilitate CO_2 capture, but these have not yet been proven. A variety of advanced gas separation methods are being developed by national and international R&D efforts to selectively and more cost-effectively remove CO_2 from flue gas and other gas streams. However, large-scale demonstrations of CO_2 capture and sequestration at the 100 MW scale are needed before such systems can be implemented on a large scale.

---

[2]Oxyfuel combustion systems involve the combustion of pulverized coal in a mixture of oxygen and recirculated flue gas in order to reduce the net volume of flue gases from the process and to substantially increase the concentration of CO_2 in the flue gases compared to the normal pulverized coal combustion in air (CCSD, 2007).

FEDERAL COAL UTILIZATION R&D PROGRAMS

Federal R&D programs related to coal utilization are located primarily in the U.S. Department of Energy (DOE). A number of smaller programs at the U.S. Environmental Protection Agency (EPA) also are related to coal utilization.

DOE Coal Utilization R&D Programs

The DOE Office of Fossil Energy (DOE-FE) is responsible for pursuing research, development, and demonstration (RD&D) efforts to make coal power plants less expensive, cleaner, and more efficient. These efforts are administered through the National Energy Technology Laboratory (NETL) and are focused on developing cost-effective coal use technologies and environmental controls that have the potential to yield near-zero emissions (NETL, 2006a). The NETL RD&D programs include Gasification, Advanced Turbines, Carbon Sequestration, Innovations for Existing Plants (IEP), Fuel Cells, Fuels, and Advanced Research. Within the IEP program, NETL is supporting research on emissions control for mercury, advanced NO_x, and particulate matter; utilization of coal by-products; and air and water quality.

NETL's clean coal demonstration projects have the goal of demonstrating and deploying advanced clean coal technologies that, in DOE's words, will "benefit the environment, enhance electricity reliability, bolster energy security, and help to ensure an affordable supply of electricity" (DOE, 2004). Among the projects are two government-industry partnership programs—the Clean Coal Power Initiative (CCPI) and the Power Plant Improvement Initiative (PPII). DOE provides up to 50 percent of the funding for each project, and the industrial partners involved contribute the remainder. The programs have somewhat different goals:

- The CCPI program was designed to support the demonstration of a range of promising technologies with potential to efficiently and reliably generate electric power with minimum adverse impact on the environment. This 10-year program was established in 2001 to increase investments in clean coal technology.
- The PPII program was a one-time program conducted in 2000 to implement commercial-scale demonstration of clean coal technologies at existing and new electric power generation facilities, with the objective of demonstrating higher efficiencies, lower emissions, improved economics, and enhanced system performance. Improving by-product utilization was also one of the areas of focus in the PPII program.

NETL's Coal and Power Systems Program includes the Gasification Technologies Program and the FutureGen Project, as well as a component that sponsors advanced research. Research projects are under way in gasification, advanced combustion, turbine and heat engine technologies, and carbon sequestration:

- The Gasification Technologies Program emphasizes R&D on gas cleaning and conditioning. Activities include research on (1) a two-stage process for the removal of H_2S, trace metals, HCl, and particulates; (2) membrane processes for control of H_2S, mercury, and CO_2; and (3) sorbents for NH_3 control.
- FutureGen is a planned 275 MW prototype plant that will use coal gasification and other advanced technologies to produce electricity, hydrogen, and other products, with near-zero emissions at high efficiency levels. The facility will serve as a large-scale engineering laboratory for testing new clean power, CCS, and coal-to-hydrogen technologies. The plant is expected to begin operating in the 2012 time frame.
- One of DOE's fastest-growing programs in recent years has been the Carbon Sequestration Program, which focuses on CO_2 capture and storage technologies with high potential for reducing greenhouse gas emissions. The initial goal was to develop instrumentation and measurement protocols for direct sequestration in geological formations and for indirect sequestration in forests and soils. Other program goals are to begin demonstration of large-scale carbon storage options by 2008, and to develop—to the point of commercial deployment—systems for advanced indirect sequestration of greenhouse gases, also by 2008. Biological and chemical processes that convert CO_2 to solid materials also are being investigated as means of sequestering CO_2, and oceanic and terrestrial carbon sequestration has been studied (NRC, 2003b; DOE, 2007c). However, the principal focus of the DOE program is on geological sequestration, which is the only long-term storage option that has been demonstrated at a commercial scale—three large facilities are already in operation worldwide, and a number of smaller injection sites operate for research and development purposes (IPCC, 2005). In terms of capture technology, DOE's target is to develop—to the point of commercial deployment—systems for direct capture and sequestration of greenhouse gases and pollutant emissions from fossil fuel conversion processes by 2015. The goal is for these systems to have near-zero emissions and approach a no-net-cost increase for energy services (DOE, 2007a).
- DOE's NETL also sponsors basic research through its Advanced Research Program. This program supports research in a variety of areas and strives to bridge the gap between basic research and the development of new systems capable of improving the supply and utilization of fossil energy resources.

DOE also manages several programs that focus on technologies to facilitate the production of hydrogen from coal for use in fuel cells and other systems. Two possible coal-based hydrogen production scenarios are considered in the program: (1) the production of hydrogen alone or in combination with electricity, and (2) the production of high-hydrogen-content liquid fuels that can be transported through the existing infrastructure to sites where the fuels would be extracted and stored near the customer.

Currently absent from the portfolio of DOE-sponsored projects are any full-scale demonstration projects of integrated CO_2 capture and storage at a modern coal-based power plant. The largest planned project is the 275 MW FutureGen plant noted above, which is classified as a research project rather than a demonstration project. Nonetheless, FutureGen will provide valuable information on the design and operation of IGCC plants with carbon capture and sequestration (CCS). In addition, DOE also recently announced its intention to support seven large-scale demonstrations of CO_2 sequestration in Phase III of its Carbon Sequestration Regional Partnerships program. As of February 2007, there were no plans for full-scale demonstrations of other CCS technologies, particularly post-combustion capture systems applicable to the current and future fleet of PC power plants in the United States. Nor are there any planned U.S. demonstrations of oxyfuel combustion systems for CO_2 capture at PC plants, although such projects are underway in other countries.[3] In recognition of the need for additional full-scale demonstrations of CCS, the Coal Utilization Research Council (CURC) and the Electric Power Research Institute (EPRI) recently called for a $6.7 billion program of technology demonstrations, including demonstrations of CO_2 capture at both combustion- and gasification-based facilities, plus a $900 million initiative for three demonstration projects and expanded R&D for CO_2 sequestration (Burke and Carter, 2006).

While preliminary studies indicate that there are many geological formations in the United States that are potentially capable of storing large quantities of CO_2 (DOE, 2006), more detailed assessments are needed to identify suitable storage sites, quantify their capacities to store CO_2, assess and quantify carbon dioxide migration and leakage rates at the storage sites, understand the environmental impacts of storage, and make the sites available (see Box 6.3). To a large degree, these important research areas can and are being addressed by participants in the current DOE Carbon Sequestration Program, including the seven regional partnerships in different parts of the country.

EPA R&D Programs Associated with Coal Utilization

Although primarily focused on regulation, the U.S. Environmental Protection Agency (EPA) does support limited research through the Office of Research and Development. While it does not have a program specifically focused on coal utilization, a number of EPA activities are directly or indirectly related to coal use, for example:

---

[3]In March 2007, American Electric Power (AEP) issued a press release to announce its intention to install post-combustion carbon capture technology on two coal-fired power plants. The announcement indicated that AEP plans to have the technology in service at a plant in 2012-2015 (AEP, 2007).

BOX 6.3
Geological Sequestration of CO_2—Resources,
Reserves, and Characterization

Geologic sequestration of CO_2 will require specific geologic strata in which the CO_2 can be placed and where it will remain permanently. As such, these geologic units are "resources," but in a different sense than mineral or energy resources. Sequestration resources have the ability to host CO_2 at some time in the future, whereas mineral and energy resources are currently hosts to their desired elements.

The primary targets for geologic sequestration of CO_2 are petroleum reservoirs, saline reservoirs, and deep, unminable coal seams. The first two are widely distributed in sedimentary basins and have the potential to provide storage for large quantities of CO_2. Coal seams are more limited in their ability to sequester CO_2 on a worldwide basis, but individual coal seams may be an attractive storage target because in some settings coal may sequester a greater mass of CO_2 per unit volume than the other reservoirs, and there is also the potential of recovering methane from coal during the sequestration process. Recently, researchers have found it useful to characterize the geologic sequestration process by the types of trapping mechanisms that can occur in the subsurface—structural and stratigraphic trapping, residual gas trapping, dissolution trapping, mineral trapping, hydrodynamic trapping, and coal adsorption (Bradshaw et al., 2005).

Reliable estimates of the amount of geologic storage of CO_2 that can be located in any region depend on the ability to determine accurate capacity estimates for the "reservoirs" in the various types of storage sites. Current estimates for world storage capacity range over four orders of magnitude, indicating that these estimates are less than satisfactory. Many of the estimates used overly simplistic methods and were based only on the total surface area of a sedimentary basin, using an assumed average thickness of reservoir rock and an average porosity (Bradshaw et al., 2005).

- The Particulate Matter (PM) Research Program is focused primarily on understanding how fine particles are emitted into the atmosphere, how they are formed in the atmosphere from gaseous pollutants, and how they are transported. Research is also under way to better understand the attributes of particles that cause adverse health effects, to identify those who may be most susceptible to these effects, and understand how people are exposed to PM air pollution.
- The EPA has undertaken a range of research related to mercury emissions, including the development of protocols for verifying continuous air emission monitors used to measure total and chemically speciated mercury in source emissions, and the compilation of speciated (elemental, oxidized, and particulate) mercury emissions data from coal-fired utility units to estimate mercury emissions nationwide.

Reliable estimates of CO_2 storage capacity are required by national, regional, and local governments, as well as by the emerging sequestration industry. Estimation of CO_2 storage capacity requires a detailed knowledge of the characteristics of the potential reservoir (Bradshaw et al., 2005, 2006; Brennan and Burruss, 2006). Although there are no currently accepted guidelines for classifying CO_2 storage "resources" and "reserves," a classification system for CO_2 storage capacity has been proposed by Frailey et al. (2006).

Bradshaw et al. (2005) identified four gaps that require additional research into standards for measurement of CO_2 storage capacity:

- Identification of clear and accepted definitions that are meaningful across a range of geoscience disciplines, including geology, reservoir engineering, and hydrology
- Establishment of consistent and accepted methodologies and guidelines for capacity estimation
- Establishment and documentation of appropriate constraints for assessments, especially for the technical (geological and reservoir engineering) data
- Establishing reporting practices for storage capacities that are on a par with modern practices in the other resource industries

The DOE's Office of Fossil Energy is the lead federal entity for development of carbon sequestration technology in the United States. The program, administered by DOE's National Energy Technology Laboratory, is extensive—$67 million was enacted in FY 2006 and nearly $74 million was requested for FY 2007. This program included support for a high-level overview of potential geologic sequestration sites in the United States and Canada (DOE, 2007c). A large number of state agencies, universities, and private companies are involved in the program through seven regional partnerships as well through a variety of other projects funded by the program.

EPA also has regulatory responsibility for waste disposal that involves underground injection of hazardous and nonhazardous wastes. Injection of CO_2 for geological sequestration currently falls within the purview of the EPA Underground Injection Control (UIC) program, whose primary purpose is the protection of drinking water supplies. CO_2 injection for enhanced oil recovery (EOR) is currently permitted under the UIC, and EPA has recently formulated draft guidance to permit pilot geological sequestration projects involving injections into deep saline formations. However, the regulatory structure and requirements applicable to future large-scale CCS programs are yet to be developed, although such issues are receiving considerable attention from a variety of interest groups in the United States and elsewhere (NETL, 2006b).

FINDINGS AND RECOMMENDATION—COAL UTILIZATION

The dominant use of coal in the United States today is for electric power generation based on the combustion of pulverized coal. As a result of regulatory programs over the past two decades, substantial progress has been made toward the control of emissions of air pollutants and other wastes that can harm human health (EPA, 2007). Federal R&D programs concerned with coal use are focused on developing more cost-effective technologies to enable coal-based power plants to meet environmental constraints.

Emissions of CO_2 from coal-based power plants are not currently subject to regulation or controls. However, low-emission coal-based power plants equipped with technologies for the capture and geological sequestration of CO_2 are projected to be developed to the point of commercial readiness by 2015 as part of DOE's Carbon Sequestration Program. Currently, however, there are no full-scale demonstration projects of CO_2 capture and storage at a large, modern coal-based power plant.

The following principal findings have resulted from this brief review of coal utilization issues:

• Potential regulatory requirements to further reduce emissions of NO_x, SO_2, mercury, and particulate matter in the future are not expected to significantly limit the overall use of coal in the next several decades. However, future emission control requirements for these regulated air pollutants could result in changed preferences for particular types of coal, depending on the nature of future regulations.

• Decisions to invest or not invest in coal-based power plants will strongly influence future coal use, and this will depend in large part on both the timing and the magnitude of any future constraints on CO_2 emissions, as well as on the demonstrated commercial readiness of technologies to significantly reduce CO_2 emissions from coal-based power plants and other energy conversion processes.

• Detailed assessments are needed to identify and characterize potential geological formations in the United States that are capable of sequestering large quantities of CO_2; to quantify their storage capacities; to assess migration and leakage rates; and to understand the economic, legal, and environmental impacts of storage on both near-term and long-term time scales. Such geologic sequestration sites should be considered "resources," and categorized and described in the same way that conventional mineral or energy resources are assessed.

Recommendation: The U.S. Geological Survey (USGS) should play a lead role in identifying, characterizing, and cataloguing the CO_2 sequestration capacity of potential geologic sequestration resources.

The USGS has expertise in coordinating nationwide assessments of oil and gas, minerals, and coal, and a history of developing consensus in resource and reserve terminology. It would be appropriate for the USGS to have the same lead role in a national assessment of the nation's geologic carbon sequestration resource that it currently has with regards to coal and petroleum resources. The committee estimates that approximately $10 million per year for five years will be required for this activity.[4] There should be close cooperation and coordination among the USGS, the Carbon Sequestration Program managed by DOE-FE, and the states involved in the Carbon Sequestration Regional Partnerships.

---

[4]In March, 2007, a bill was introduced to the U.S. Senate (S.731) to authorize funding for the USGS to carry out a national assessment of geological storage capacity for carbon dioxide.

7

Coal Research Needs and Priorities

The task of this committee (Box 1.2) was to broadly examine coal research, technology, and resource assessments, recognizing that they are essential components of an integrated roadmap for our nation's future energy needs. The intent of this analysis was to allow policy makers to gauge the success of past research activities, gain a clearer understanding of the research presently being undertaken throughout the entire coal cycle, and provide updated and expanded information as the basis for improved prioritization of investment within the coal sector. By also examining critical gaps in research and technology, and the potential impacts of key policy developments, this study was intended to offer a more complete picture of the role of coal in the U.S. energy mix and provide the basis for more informed development of a national energy strategy.

The United States today relies on coal for nearly a quarter of all domestic energy supplies, with most of that coal used to generate more than half of the nation's electricity. Projections by the U.S. Energy Information Administration (EIA) and others foresee increasing demands for electricity over the coming decades and, with it, increasing demands for coal. Recent price increases for natural gas and petroleum-based transportation fuels have made the outlook for coal increasingly bullish, not only for power generation but also for conversion of coal to substitute natural gas and liquid fuels for transportation. The widely cited projections from the EIA's 2006 *Annual Energy Outlook* call for a 50 percent increase in U.S. coal use by 2030 in the reference case scenario and an approximately 70 percent increase in the "high-oil-and-gas-price" scenario. Projections such as these provided much of the underlying motivation for the present study, to enable any potential constraints and research and development (R&D) needs that may be required to meet increasing demands for coal to be addressed in a timely manner.

Perhaps less widely known is that the EIA projections in its *Annual Energy Outlook* assume no change in current U.S. laws and regulations affecting energy supplies or demand over the next 25 years. Therefore, while the *Annual Energy Outlook* examines the implications of alternative assumptions regarding energy prices, technology costs, economic growth, and other parameters in its forecasting models, the EIA is precluded from analyzing alternative policy scenarios in that report. However, as a result of specific requests from Congress, EIA models have been used to analyze policy proposals that would require reduced emissions of CO_2 and other greenhouse gases. These analyses reveal a very different outlook for coal. In most of these scenarios, future growth in coal use is significantly curtailed and in some cases even falls below 2004 levels by 2030. However, other scenarios and models project significant increases in coal use even with stringent CO_2 constraints, provided that carbon capture and sequestration (CCS) technology is available to sequester CO_2 in geological formations. Overall, the review of current domestic and international coal use forecasts presented in Chapter 2 reveals that potential future constraints on greenhouse gas (especially CO_2) emissions, and the technical and economic feasibility of CO_2 control measures, are the dominant issues affecting the outlook for the future of coal use over the next 25 years and beyond. The difficulty with predicting the prices and availability of alternative energy sources for electric power generation provides additional uncertainty. The committee explicitly found the following:

- Over the next 10 to 15 years (until about 2020), coal production and use in the United States are projected to range from about 25 percent above to about 15 percent below 2004 levels, depending on economic conditions and environmental policies. By 2030, the range of projected coal use in the United States broadens considerably, from about 70 percent above to 50 percent below current levels. The higher values reflect scenarios with high oil and gas prices and no restrictions on carbon emissions. The lower values reflect scenarios with relatively strict limits on U.S. CO_2 emissions, which cause coal use with sequestration to be more costly compared to other options for power generation.
- At present, coal imports and exports represent small fractions of total U.S. coal production and use. Projections indicate that imports and exports are expected to remain relatively small.
- Globally, the largest tonnage increases in coal use are expected in the emerging economies of China and India. Much smaller tonnage growth is projected in the rest of the world, although relative growth rates are projected to be high in several other countries. Again, however, there is great uncertainty in global coal use projections, especially beyond about 2020.

It is with the above outlook in mind that the committee undertook an assessment of major needs for coal-related research and development. In accordance with directions from the congressional framers of this study, the primary focus is on "upstream" components of the overall coal fuel cycle. The remainder of this chapter notes two important societal issues—community impacts and workforce demographics—that cut across the coal fuel cycle and then describes current federal support for coal-related R&D. The findings and recommendations from earlier chapters are repeated here to provide the context for a federal R&D investment strategy, coordinated among federal agencies, coal-producing states, and the coal industry, for upstream components of the coal fuel cycle.

SOCIETAL ISSUES AND COAL

Two societal issues that occur across multiple components of the coal fuel cycle are (1) the community impacts of coal mining, transport, and utilization; and (2) the education and training of the mining workforce and the academic research and teaching profession.

Community Impacts

Coal mining has both beneficial and adverse effects at all levels, from individual communities to the nation as a whole, but it is the local communities that are at the forefront of these effects. A number of socioeconomic issues exist in some older mining districts that reflect some of the unique aspects of mining as a land use. The impacts of mining on the safety and general welfare of coal communities can include mine drainage, mine fires, waste piles, ground movements (subsidence), and hydrological impacts. An additional concern in new mining districts, such as those in the West, is that the rapid development of sparsely populated areas will produce a sharply increased demand for infrastructure and community facilities that may be very difficult or cost-prohibitive to meet. Beneficial impacts are realized during the productive life of a coal mining operation, and great progress has been made over the years in minimizing adverse impacts. Maintaining a healthy community following mine closure requires deliberate planning to develop new opportunities for the community. The key to establishing sustainable communities is for both industry and community participants to cooperate to develop guidelines, practices, and reporting mechanisms that promote sustainable development (NRC, 1996). The development and adoption of these procedures would benefit from active research programs that lead to case studies of positive post-mining community development.

Coal transportation, especially by rail and truck, also affects the communities through which the coal passes. Long coal trains crossing local roads temporarily block those roads, adding traffic congestion and potentially delaying emergency responders. Heavy trucks hauling coal can damage roads and cause deaths or injuries in accidents.

Workforce and Education

Employment in the coal mining industry has been declining for more than 20 years (Watzman, 2004) (Figure 4.1). Mining workforce demographics have changed substantially, coinciding with increased production from surface mines and increased productivity of both surface and underground mines (Figure 4.4). However, for any of the projected scenarios that involve substantially increased coal production, the skewed age distribution of the existing coal mining workforce (Table 7.1) dramatically emphasizes the need for the industry to attract new miners in addition to replacing the retiring workforce. Similarly, railroads have cited changes in demographics, training requirements, and limits on the availability of qualified personnel as posing a risk to their ability to meet the demand for rail service. Low unemployment in the general economy has been cited as making it more difficult to hire new personnel for jobs on train crews that require considerable time away from home.

Consolidation of mining companies and the increasing size of mines over the past two decades have resulted in a marked decline in demand for technically trained personnel. This, coupled with declining funding for academic research on mining and mineral engineering issues, has resulted in fewer accredited programs at the technology and engineering levels and a decrease in the number of graduates and postgraduates from these programs. There is now a substantial shortage of technically trained personnel in the mining and mineral engineering

TABLE 7.1 Age Distribution of Employees in the Coal Mining Industry in 2005

Age	Number Employed	Percentage	Cumulative Percent
16-19	1,000	1.2	1.2
20-24	3,000	3.7	4.9
25-34	13,000	15.8	20.7
35-44	16,000	19.5	40.2
45-54	36,000	44.0	84.2
55-64	13,000	15.8	100.0
65+	0	0	
Total	**82,000**	**100**	

NOTE: The median age for mining employees was 46.1 years, compared with 40.7 years for the overall workforce.

SOURCE: BLS (2006).

disciplines, and this shortage will be exacerbated by any significant increases in coal production.

The limited research funding by industry and government has been a serious threat to the sustainability of postgraduate programs, and the reduction in mining research has had a significant impact on the recruitment, retention, and development of faculty in mining-related disciplines. The vast majority of the academic faculty in mining programs is at a senior level and close to retirement, posing serious succession and continuity problems. Extramural funding by federal agencies directed to universities in support of fundamental research in the earth sciences and engineering would support postgraduate programs and assist in recruiting, retaining, and developing the academic mining profession (NRC, 2004b, 2005b, 2007a).

EXISTING FEDERAL SUPPORT FOR COAL-RELATED R&D

Coal-related R&D is carried out by a range of organizations and entities—federal government agencies, state government agencies, academic institutions, coal mining companies, and equipment manufacturers. In general, the scope and motivation for research are determined by the relevance and potential impact of solutions to the problems that need to be responded to by the various entities and organizations (e.g., agencies with a primarily regulatory role support limited research focused on technical support for regulation; equipment manufacturers undertake materials research and market trends). Here, the committee focuses on existing support for research and development funding by the federal government across the coal fuel cycle, to set the context for recommendations contained in the following sections.

For this report, the committee adopted a broad interpretation of R&D to include activities that are variously described by federal agencies as pure research, applied science, technical support, pilot-scale testing, demonstration projects, and applied engineering projects. Budgets were requested for the 1995, 2000, and 2005 fiscal years so that funding trends might be discerned. The data were analyzed by the committee, and the committee's interpretations of the data were sent back to agency staff for confirmation. The committee categorized R&D budgets according to the different stages of the coal fuel cycle—resource and reserve assessment, coal mining and processing, coal mining safety and health, environmental protection and reclamation, transport of coal and coal-derived products (including electricity transmission), and coal utilization (including air emission and carbon sequestration research).

More than $538 million was spent by federal government agencies for coal-related research and technology development in 2005 (see Table 7.2). The specific coal-related roles of these agencies are described in Appendix C, together with limited descriptions of past levels of federal R&D support. Table 7.2 shows that coal-related R&D support from most offices is focused on a single R&D

TABLE 7.2 Summary of Federal Agency Support for Coal-Related Research and Technology Development in 2005

	Resource and Reserve Assessment	Coal Mining and Processing			Transport and Transmission	Coal Utilization and CCS	Total (dollars)
		Productivity and Resource Optimization	Safety and Health	Environment and Reclamation			
Department of Energy							
Energy Information Administration	15,000						15,000
Energy Efficiency and Renewable Energy		734,000[a]	110,000				844,000
Office of Fossil Energy						442,204,000	442,204,000
Office of Electricity Delivery and Energy Reliability					48,470,000[b]		48,470,000
Environmental Protection Agency							
Office of Research and Development				9,200,000[c]			9,200,000
Mine Safety and Health Administration			1,274,000[c]				1,274,000
National Institute for Occupational Safety and Health			23,006,000				23,006,000
National Science Foundation	36,000	594,000	8,000	124,000	260,000	1,807,000	2,829,000
Office of Surface Mining Reclamation and Enforcement				600,000			600,000
U.S. Geological Survey	9,700,000[d]						9,700,000
Total	**9,751,000**	**1,328,000**	**24,398,000**	**9,924,000**	**48,730,000**	**444,011,000**	**538,142,000**

NOTE: Numbers are rounded to nearest thousand.

[a]Of the coal-related projects undertaken during the 8-year history of the Mining Industry of the Future program (1999-2006), 87% addressed mining and processing issues and 13% addressed safety and health issues. The funding for 2005 was distributed according to this ratio.

[b]This figure is derived by allocating 53% of OE's R&D appropriation to coal-related activities; 53% of national electricity generation is supplied by coal-fired power plants.

[c]2006 data.

[d]This number represents what was requested by the agency, rather than the actual dollars spent on coal-related research projects.

category. The basic research supported by the National Science Foundation (NSF) is a clear exception, providing relatively limited support that extends across all categories.

The Department of Energy (DOE) supports more than 91 percent of all coal-related research and technology development, exceeding research support from all other agencies by a very wide margin (Figure 7.1). The distribution of R&D funding by category (Figure 7.2) shows that almost 92 percent of the federal agency funding is for downstream components of the coal fuel cycle, mostly coal utilization technology development and transmission research funded through DOE (Figure 7.1). Federal support for R&D activities related to all upstream aspects of the coal fuel cycle (i.e., mine worker safety and health, resource and reserve assessments, coal mining and processing, environmental protection and reclamation) accounted for less than 10 percent of the total federal investment in coal-related R&D. Federal funding in 2005 for individual components of upstream activities ranged from $24.4 million (4.5 percent) for mine worker safety and health R&D to $1.3 million (0.2 percent) for coal mining and process-ing R&D.

Overall, federal government funding for regulatory and downstream com-ponents of the coal fuel cycle over the past decade has continued at a constant level or increased, while support for upstream R&D has decreased (Table 7.3; Appendix C). Since 1995, support for resource and reserve assessments (U.S. Geological Survey [USGS] and DOE-EIA) has declined by nearly 30 percent, and support for health and safety R&D (predominantly by the National Institute

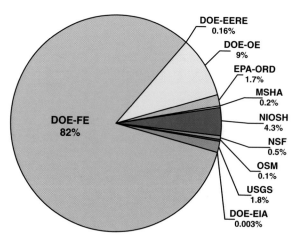

FIGURE 7.1 Distribution of funding in 2005 for coal research, technology development, and resource assessment by federal government agency or office.

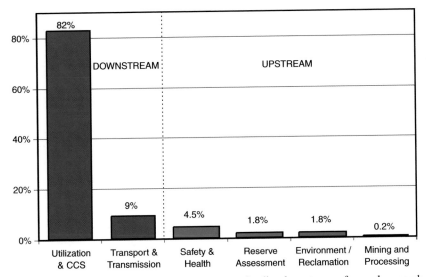

FIGURE 7.2 Distribution of federal government funding by category for coal research, technology development, and resource assessment in 2005.

TABLE 7.3 Summary of Trend Data for Federal Agency Support for Coal-Related Research and Technology Development

	Appropriation for 2005	Appropriation for 2000 and Comparison with 2005		Appropriation for 1995 and Comparison with 2005	
Department of Energy					
Energy Information Administration	15,000	17,000	−12%	64,000	−77%
Energy Efficiency and Renewable Energy	844,000	1,065,000	−21%	NA	NA
Office of Fossil Energy	442,204,000	326,543,000	+35%	355,556,000	+24%
Environmental Protection Agency—Office of Research and development	9,200,000	9,800,000	−6%	8,600,000	+7%
Mine Safety and Health Administration	1,274,000	1,435,000	−11%	1,416,000	−10%
National Institute for Occupational Safety and Health	23,006,000	20,818,000	+11%	54,283,000	−58%
Office of Surface Mining Reclamation and Enforcement	600,000	284,000	+111%	256,000	+134%
U.S. Geological Survey	9,700,000	12,476,000	−22%	13,580,000	−29%

NOTES: Numbers are rounded to nearest thousand, all in constant 2005 dollars. NSF basic research awards, because they are distributed over a range of programs from different directorates, were not suitable for trend analysis.

for Occupational Safety and Health [NIOSH] and the Mine Safety and Health Administration [MSHA]) has declined by 56 percent. During this same period, support for R&D related to downstream utilization (DOE-FE and EPA-ORD) has increased by 24 percent.

IMPROVED COAL RESOURCE, RESERVE, AND QUALITY ASSESSMENTS

Accurate and complete estimates of national reserves are needed to determine whether coal can continue to supply national electrical power needs and whether coal has the potential to replace other energy sources, such as petroleum, that may become less reliable or less secure. Two major questions are considered in Chapter 3 to assess existing estimates of the amount of usable coal:

1. Are estimates of available coal reliable, and are they good enough to allow federal policy makers to formulate coherent national energy policies?
2. Can coal reserves in the United States produce the 1.7 billion tons per year total of coal required in 2030 if the EIA reference case described in Chapter 2 becomes a reality?

The two primary federal agencies that provide resource and reserve information are the U.S. Energy Information Administration (EIA) in the Department of Energy, and the U.S. Geological Survey (USGS) in the Department of the Interior. The EIA is responsible for maintaining Demonstrated Reserve Base (DRB) data (Box 3.1), the basis for assessing and reporting U.S. coal reserves. The USGS has responsibility for mapping and characterizing the nation's coal resources, in cooperation with agencies that have land and resource management responsibilities (e.g., the Bureau of Land Management [BLM] and the Office of Surface Mining Reclamation and Enforcement [OSM]) and agencies that use USGS resource projections (e.g., the EIA). Although most coal-producing states have geological surveys that collect data to categorize their coal resources, in most cases these organizations lack the personnel and funding for comprehensive coal resource and reserve investigations; most state coal resource investigations have been undertaken in cooperation with the USGS, BLM, or OSM. For this reason, they typically only evaluate the in-place tonnage and do not estimate recoverability—this has been largely left to the USGS and the EIA. Mining companies generate detailed reserve estimates for the coal they control or are interested in obtaining. Companies consider these data to be proprietary, and as a consequence they are rarely available for government resource and reserve studies except for the reserve estimates that have to be reported at operating mines. In assessing existing estimates of available coal and the data and methodologies used to derive these estimates, the committee came to the following conclusions:

- The United States is endowed with a vast amount of coal. Despite significant uncertainties in generating reliable estimates of the nation's coal resources and reserves, there are sufficient economically minable reserves to meet anticipated needs through 2030. Further into the future, there is probably sufficient coal to meet the nation's needs for more than 100 years at current rates of consumption. However, it is not possible to confirm the often-quoted suggestion that there is a sufficient supply of coal for the next 250 years. A combination of increased rates of production with more detailed reserve analyses that take into account location, quality, recoverability, and transportation issues may substantially reduce the estimated number of years supply. Because there are no statistical measures to reflect the uncertainty of the nation's estimated recoverable reserves, future policy will continue to be developed in the absence of accurate estimates until more detailed reserve analyses—which take into account the full suite of geographical, geological, economic, legal, and environmental characteristics—are completed.
- The DRB and the Estimated Recoverable Reserves (ERR), the most cited estimates of coal resources and reserves, are based upon methods for estimating resources and reserves that have not been reviewed or revised since their inception in 1974. Much of the input data for the DRB and ERR also date from the early 1970s. These methods and data are inadequate for informed decision making. New data collection, in conjunction with modern mapping and database technologies that have been proven to be effective in limited areas, could significantly improve the current system of determining the DRB and ERR.
- Coal quality is an important parameter that significantly affects the cost of coal mining, beneficiation, transportation, utilization, and waste disposal, as well as its sale value. Coal quality also has substantial impacts on the environment and human health. The USGS coal quality database is largely of only historic value because relatively few coal quality data have been generated in recent years.

Recommendation: A coordinated federal-state-industry initiative to determine the magnitude and characteristics of the nation's recoverable coal reserves, using modern mapping, coal characterization, and database technologies, should be instituted with the goal of providing policy makers with a comprehensive accounting of national coal reserves within 10 years.

The U.S. Geological Survey already undertakes limited programs that apply modern methods to basin-scale coal reserve and quality assessments. The USGS also has the experience of working with states to develop modern protocols and standards for geological mapping at a national scale through its coordinating role in the National Cooperative Geologic Mapping Program. The USGS should be funded to work with states, the coal industry, and other federal offices to quantify and characterize the nation's coal reserves. The committee estimates that a comprehensive accounting of national coal reserves would require additional funding of approximately $10 million per year.

RESEARCH TO SUPPORT COAL MINING AND PROCESSING

The transition of coal from resource to reserve requires that the coal is minable from both a technical and an economic standpoint, so resource assessment processes must be closely linked to mining processes. Research and development activities offer the potential to solve the range of challenges associated with the more difficult mining conditions of the future, thereby maximizing the nation's coal reserves.

Coal mining and processing involve a series of sequential operations: (1) exploration of a potentially economic coal seam to assess minable reserves, environmental issues, marketable reserves, potential markets, and permitting risks; (2) analysis and selection of a mining plan; (3) securing the markets; (4) developing the mine; (5) extracting the coal; (6) processing the coal if necessary; and (7) decommissioning the mine and releasing the property for future post mining uses. These activities, outlined in Chapter 4 (and amplified in Appendix E), result in a range of mining and processing challenges that in most cases already exist today, but are likely to become more pronounced in the mines of the future. As near-surface coal deposits are depleted, surface operations will mine deeper seams that require increased stripping ratios and multiple benches. Similarly, underground coal mines will have to access seams that are deeper, thinner, or thicker, generally with higher gas content and potentially more difficulties in control of the associated strata (i.e., ground control). In some cases, overlying seams will already have been mined or, to meet increased production, multiple seams may have to be mined simultaneously. These more difficult mining conditions will require improved methods to protect the health and safety of mine workers, to improve environmental management of mined lands and wastes, and to provide higher rates of resource recovery and mine productivity.

Improved Mine Worker Health and Safety

Factors that increase health and safety risks to the coal mining workforce include the introduction of new equipment and systems, commencement of mining in virgin areas, infusion of new workers, and the mining of multiple seams—seams that are thinner, thicker, or deeper than those customarily mined at present, and new seams that underlie or overlie previously mined-out seams. Additional risk factors that are likely to apply in the deeper mines of the future are the potential hazards related to methane control, dust control, ground control, ignition sources, fires, and explosions. All of these factors are likely to apply to some degree in future mines, irrespective of whether the higher production scenarios that are foreseen in some forecasts eventuate. If they do eventuate, these risks are likely to become even more pronounced. The committee found that there are major knowledge gaps and technology needs in the areas of escape and survival, communications systems, and emergency preparedness and rescue. In addition, greater understanding and better prediction of strata control to prevent unanticipated roof collapse are essential for maintaining and improving worker safety. The funding context is that federal support for health and safety research significantly decreased about a decade ago, and has essentially remained constant since that time.

> **Recommendation: Health and safety research and development should be expanded to anticipate increased hazards in future coal mines.** These R&D efforts should emphasize improved methane control, improved mine ventilation, improved roof control, reduced repetitive and traumatic injuries, reduced respiratory diseases, improved escape and rescue procedures, improved communications systems, and research to reduce explosions and fires. This should be coupled with improved training of the mining workforce in all aspects of mine safety. R&D should also be directed toward lowering the exposure of mine workers to hazardous conditions, particularly through expanded use of remote sensing and the automation of mining operations.

Most mining health and safety research by the federal government is carried out by the Mining Program at the National Institute for Occupational Safety and Health. Technology-related activities within the Mine Safety and Health Administration are limited to technical support and training services for its personnel and those from the mining industry. With NIOSH carrying out the research needed to improve mine safety and support MSHA's regulatory role, these two agencies play a vital role in coal mine health and safety. The committee estimates that the enhanced health and safety program proposed here will require annual R&D funding of approximately $60 million, and recommends that NIOSH continue as the lead agency with enhanced coordination with MSHA and industry.

Improved Environmental Protection

As mining extracts coal from deeper and operationally more difficult seams by both surface and underground methods, a range of existing environmental issues and concerns will be exacerbated, and new concerns—particularly related to greater disturbance of hydrologic systems, ground subsidence, and waste management at mines and preparation plants—are likely to arise. Inadequate understanding of post-mining strata behavior and the associated hydrologic consequences of mining in both surface and underground mines affects mine permitting, mine development, environmental mitigation, and post-mining land use, including use for waste management. Research offers considerable potential to mitigate the effects of past mining practices, particularly acid mine drainage on abandoned mine lands. However, the regulatory environment and the technical support programs administered by both state and federal agencies, and implemented by mining companies through their compliance practices, are inadequately supported by existing research programs.

> **Recommendation: Additional research is needed to mitigate the adverse environmental impacts associated with past, existing, and future coal mining and processing. Research activities should focus particularly on developing techniques to mitigate the alteration and collapse of strata overlying mined areas, to model the hydrological impacts of coal mining, to improve mine mapping and void detection, to improve the stability of spoils on steep slopes, and to improve the construction and monitoring of impoundments.**

Both the Office of Surface Mining Reclamation and Enforcement and the U.S. Environmental Protection Agency, although primarily regulatory agencies, fund limited R&D activities in support of their missions. The committee estimates that annual funding of approximately $70 million will be required to conduct the research necessary to adequately respond to the environmental impacts of past, existing, and future mining operations. The committee recommends that OSM should be the lead agency in this effort, and it should coordinate closely with related EPA and state research activities.

Improved Mine Productivity and Resource Optimization

Although evolutionary developments in technology and mining practices (primarily underground longwall mining in the East and the growth of large surface operations in the West) have resulted in a two- to threefold increase in the productivity of U.S. coal mines since the mid-1970s, production and productivity increases in recent years have been small or nonexistent as mining companies and equipment manufacturers made only incremental improvements. Over the past decade, there has been little R&D directed toward truly advanced mining technologies and at present, only 0.2 percent of total federal coal-related R&D funding is directed toward development of the advanced mining technologies and practices that are necessary to optimize utilization of the nation's coal resource. Small percentage increases in coal recovery through improved mining and coal preparation processes have the potential to significantly expand economically recoverable reserves of both eastern and western coals. The development of these technologies, increasingly needed as coal reserve quality decreases over time, will help to maximize utilization of the nation's coal resource.

The global transfer of coal mining and processing technology within the industry is facilitated by international equipment manufacturers, who work closely with suppliers and the larger mining clients on evolutionary product developments. However, there is little evidence of the efficient transfer of technologies from outside the mining industry. This is at least partly due to the relatively small market that the coal mining industry represents to potential technology suppliers and the scarcity of coal mining research at academic institutions and national laboratories.

> **Recommendation: There should be renewed support for advanced coal mining and processing research and development to optimize use of the nation's coal resources by increasing the amount of coal that is economically minable through technological advances that accommodate health, safety, and environmental requirements. The focus of this R&D should be on increased integration of modern technology in the extraction and processing phases of coal production, with particular emphasis on emerging advances in materials, sensors, and controls; monitoring; and automated mining systems.**

Research to develop advanced mining technologies requires not only cooperation among relevant federal agencies, but also participation by academic institutions, as well as funding, guidance, and technology transfer by industry. The committee estimates that advanced coal mining and processing R&D will require approximately $60 million per year and recommends that this funding be comprised of $30 million in federal support, with cost sharing from non-federal sources. The DOE Office of Fossil Energy should be the lead federal agency, and should coordinate with NSF, OSM, NIOSH, academic institutions, and the coal industry to ensure that all research activities carefully consider the environmental, reclamation, and health and safety aspects of coal mining.

TRANSPORT OF COAL AND COAL PRODUCTS

With the electric power sector accounting for more than 90 percent of U.S. coal use, transportation of coal to the more than 600 coal-burning power plant sites in the nation is especially important. Of these plants, rail transportation serves approximately 58 percent, waterborne transportation serves 17 percent, trucks serve 10 percent, 12 percent are served by multiple modes of transportation (primarily rail and barge), and 3 percent are minemouth plants with conveyor systems. In 2004, more than 85 percent of coal shipments were delivered to consumers by either rail (684 million tons), truck (129 million tons), or water (98 million tons). One-third of all coal delivered to power plants is subject to at least one transloading along the transportation chain.

Growth in the use of coal depends on having sufficient capacity to deliver increasing amounts of coal reliably and at reasonable prices to an end user. The capacity, reliability, and price of rail transportation—the dominant mode of coal transport—depend largely on the supply and demand for rail transportation, as well as on prevailing business practices, the investment climate, and the nature of regulatory oversight of the railroad industry. The capacity, reliability, and price of rail transportation of coal depend to a far lesser degree upon research and development. Reliable and sufficient waterborne transportation—the second most prevalent method of coal transport—depends on the construction and maintenance of waterway infrastructures, especially lock-and-dam infrastructure and port capacity.

Much of the nation's coal-fired electric generating capacity is located at some distance from the urbanized areas that have the largest and most concentrated demands for electricity. Projections of higher coal use depend on sufficient capacity to transmit electricity from coal-based power plants to such areas reliably and at a reasonable cost. Conversely, the projected increases in coal use will diminish if these high-demand areas satisfy much of their growing demand for electricity not by expanding their ability to import electricity from other regions, but by a combination of energy efficiency, demand response, and local electric generation from sources other than coal.

The coal transportation and electric transmission systems are large and complex networks in which localized disruptions can have severe and widespread impacts. Weather and other natural phenomena, as well as societal factors such as sabotage and terrorism, impose a range of risks on these systems. These characteristics make it difficult to guarantee that there will be sufficient capacity to transport coal or coal-based energy (primarily electricity) reliably and cost-effectively to the various end users, particularly in light of scenarios that predict substantially increased coal use. Research is needed to better understand the factors that control these large and complex networks to minimize the risks of cascading system disruptions.

RESEARCH TO SUPPORT COAL UTILIZATION

Although the committee's task was broad and encompassed the entire coal fuel cycle, the congressional framers of this study requested that it focus primarily on the upstream aspects of the coal fuel cycle. Accordingly, only a brief overview of coal use technologies and related issues is presented in Chapter 6, with greater emphasis on describing the factors associated with coal use that are likely to impose constraints on future demands for coal.

The analysis presented in Chapter 2 indicates overwhelmingly that the environmental impacts of coal use—especially carbon dioxide emissions associated with global climate change—pose the greatest potential constraint on future coal utilization. The committee found the following:

• Potential regulatory requirements to further reduce emissions of NO_x, SO_2, mercury, and particulate matter in the future are not expected to significantly limit the overall use of coal in the next several decades. However, future emission control requirements for these regulated air pollutants could result in changed preferences for particular types of coal, depending on the nature of future regulations.

- Decisions to invest or not invest in coal-based power plants will strongly influence future coal use, and will depend in large part on the timing and magnitude of any future constraints on CO_2 emissions.

- Large-scale demonstrations of carbon management technologies—especially carbon capture and sequestration—are needed to prove the commercial readiness of technologies to significantly reduce CO_2 emissions from coal-based power plants and other energy conversion processes.

- Detailed assessments are needed to identify and characterize geological formations in the United States that are potentially capable of sequestering large quantities of CO_2; to quantify their storage capacities; to assess migration and leakage rates; and to understand the economic, legal, and environmental impacts of storage on both near-term and long-term time scales. Such geologic sequestration sites should be considered "resources," and categorized and described in the same way that conventional mineral or energy resources are assessed.

Recommendation: The U.S. Geological Survey (USGS) should play a lead role in identifying, characterizing, and cataloguing the CO_2 sequestration capacity of potential geologic sequestration resources.

The USGS has expertise in coordinating nationwide assessments of oil and gas, minerals, and coal, and a history of developing consensus in resource and reserve terminology. It would be appropriate for the USGS to have the same lead role in an assessment of the nation's geologic carbon sequestration resource that it currently has with regards to coal and petroleum resources. The committee estimates that approximately $10 million per year for five years will be required for this activity. There should be close cooperation and coordination among the USGS, the Carbon Sequestration Program managed by DOE's Office of Fossil Energy, and the states involved in the Regional Carbon Sequestration Partnerships.

COORDINATION OF COAL-RELATED R&D
BY FEDERAL AGENCIES

One component of this study was the specific requirement for the committee to evaluate whether a broad-based, coordinated, multiagency coal research and development program is required and, if so, to examine options for supporting and implementing such a program (see Box 1.1). To respond to this aspect of its charge, the committee carefully considered existing R&D programs and assessed the extent of—and opportunities for—coordination of coal-related research among the agencies, in the context of current federal funding across the coal fuel cycle

(Table 7.2). Programmatic models for R&D support that were considered by the committee included the Australian Cooperative Research Center for Mining (CRCMining) and the Australian Coal Association Research Program (ACARP) (see Box 4.3), the DOE-EERE Mining Industry of the Future program, as well as the existing coal-related research programs in federal agencies. The committee also considered coal-related R&D support provided by states, the coal industry, and equipment manufacturers but did not attempt an exhaustive compilation of these non-federal activities.

There are numerous applied research areas, primarily focused on incremental technology development, for which federal involvement is neither appropriate nor required and where the coal production industry should and does provide support. For some areas, such as ensuring that a well-trained workforce is available to meet the nation's mining and mining education requirements, federal involvement can effectively complement industry activities. There are other areas of coal-related R&D where the federal government has a primary role—for example, to establish the quantity and quality of the nation's coal reserves, to facilitate and catalyze revolutionary (rather than incremental) technology development, to safeguard the health and safety of mine workers, and to protect the environment during future mining and processing and mitigate existing environmental problems arising from past mining practices. It is also a federal responsibility to provide funding for the R&D required to support the government's regulatory role.

In considering options for R&D support, it is clear that the responsibilities and capabilities of the relevant federal agencies span a wide range. MSHA and OSM are regulatory agencies with, particularly in the case of MSHA, limited statutory authorization to conduct R&D. NIOSH and DOE-FE have well-established R&D facilities and programs, but with distinctly different missions and responsibilities. EPA has both regulatory and R&D functions, and NSF is focused solely on basic research. USGS and EIA have primarily research roles, with information-gathering and dissemination responsibilities that are quite dissimilar to those of other agencies. When considering research activities in agencies that primarily have regulatory roles, there is also the potential for reluctance by industry to reveal problems that might be appropriate targets for research solution to offices that might penalize them for having the problem. As an additional complication, these various agencies and offices are administered under a number of departments and their funding is controlled by different congressional committees.

The committee considered the recent and past history of coal-related R&D resulting from interagency cooperation (e.g., NIOSH and MSHA, USGS and OSM, USGS and EIA), as well as the R&D produced by the U.S. Bureau of Mines (USBM) (see NTIS, 2007), with its overarching mandate, prior to its demise in 1994. After considering the diverse missions and programmatic activities of the relevant agencies, the committee concluded that an attempt to consolidate all coal-related R&D into a single broad-based agency or office would be impractical at best, and would be unlikely to provide the optimum basis for a national strategy for coal R&D and resource assessments. Ultimately, the committee considered that encouraging considerably increased interagency cooperation would be more likely to achieve the goal of improved delivery of R&D to support the nation's use of its coal resource.

The choice of which model for R&D support to apply is likely to differ across agencies, with the Australian CRCMining, DOE-EERE Mining IOF program, or DOE-FE Regional Carbon Sequestration Partnerships all constituting successful models. Whatever approach an agency takes in fulfilling its R&D mission, it should make explicit provision for a high degree of accountability to ensure that the research activities are relevant to real-world needs and opportunities and are delivering their promised outcomes. This accountability process should include the publication of performance metrics as well as periodic independent external peer reviews. Programs that provide funding for extramural research that leverage government funding with non-federal dollars provide a high level of accountability. A key feature of the performance review process should be an assessment of the degree to which the agency is cooperating with other agencies as recommended in this report.

In summary, the committee found the following:

- Improved interagency cooperation and coordination of many aspects of coal-related R&D will be required to effectively respond to the full spectrum of health and safety, environmental, resource recovery, and manpower issues. Coordination should acknowledge the differing responsibilities and ranges of expertise of the individual agencies.

- Many coal-related R&D issues cut across state and regional boundaries, and a coordinated national approach—led by federal agencies and involving the states, mining companies, and universities—provides the highest likelihood of successful and effective outcomes. The Regional Carbon Sequestration Partnership's program, administered by DOE's Office of Fossil Energy, provides a model for coordinated R&D involving federal and state government agencies, industry, and national laboratories.

- The committee does not recommend a single "mega-agency" approach as the basis for coordinated coal-related R&D. Instead, the committee has identified areas where improved interagency coordination to address specific R&D opportunities and challenges could be better implemented through cooperation among two or more federal entities in R&D partnerships, with involvement of non-federal bodies as appropriate. These aspects are presented in more detail above, and summarized here together with estimates for the additional funding that will be required to support these partnerships[1] (Table 7.4).

---

[1]Rather than make recommendations concerning the distribution of recommended funding between the participants in these partnerships, the committee recognized that this would occur as part of the congressional appropriations process.

TABLE 7.4 Summary of FY 2005 and Proposed Additional Funding for Coal-Related R&D at Federal Agencies (millions of dollars per year)

	FY 2005 Funding	Proposed New Funding	Total Proposed Funding
Resource and reserve assessments and characterization[a]	10	20	30
Improved mine worker health and safety[b]	25	35	60
Environmental protection and reclamation[b]	10	60	70
Improved mining productivity and resource optimization[b]	1	29	30
Total	**46**	**144**	**190**

NOTE: FY 2005 figures are rounded to nearest million for easier comparison with proposed funding levels (unrounded figures for FY 2005 funding are presented in Table 7.2)

[a]Described in Chapters 3 and 6. Amounts do not include funding for the DOE-FE Carbon Sequestration Program, which supports a range of sequestration research and demonstration activities that include geologic sequestration site characterizations.
[b]Described in Chapter 4.

• The USGS should lead a program, in cooperation with state agencies, the coal industry, and other federal offices, to quantify and characterize the nation's coal reserves. The committee estimates that this will require additional funding of approximately $10 million per year.

• NIOSH should lead an expanded health and safety program, in cooperation with MSHA and the coal industry, to address a range of specific mine safety issues (e.g., ventilation, roof control, escape and rescue, communications systems, training) and to develop improved remote sensing and automation technologies. The committee estimates that this will require additional funding of approximately $35 million per year.

• OSM should lead an expanded program, in cooperation with EPA, state agencies, and the coal industry, to mitigate adverse environmental impacts associated with past, existing, and future coal mining and processing. The committee estimates that this will require additional funding of approximately $60 million per year.

• The DOE Office of Fossil Energy should lead a program, in cooperation with NSF, OSM, NIOSH, academic institutions, and the coal industry, to develop and demonstrate advanced coal mining and processing technology. The committee estimates that this will require a total of approximately $60 million per year and recommends that this should comprise $30 million in federal support, with cost sharing from non-federal sources.

• The USGS should lead a program to identify and characterize the nation's potential CO_2 geologic sequestration resources, with close collaboration with the Carbon Sequestration Program managed by DOE-FE, and the states involved in the Regional Carbon Sequestration Partnerships. The committee estimates that this new program will require funding of approximately $10 million per year.

Coal will continue to provide a major portion of energy requirements in the United States for at least the next several decades, and it is imperative that policy makers are provided with accurate information describing the amount, location, and quality of the coal resources and reserves that will be available to fulfill these energy needs. It is also important that we extract our coal resources efficiently, safely, and in an environmentally responsible manner. A renewed focus on federal support for coal-related research, coordinated across agencies and with the active participation of the states and the industrial sector, is a critical element for each of these requirements.

References

ACAA (American Coal Ash Association), 2005. *2005 Coal Combustion Product (CCP) Production and Use Survey.* Available online at http://www.acaa-usa.org/PDF/2005_CCP_Production_and_Use_Figures_Released_by_ACAA.pdf; accessed December 2006.

AGI (American Geological Institute), 1997. *Dictionary of Mining, Mineral and Related Terms,* 2nd ed. Alexandria, Va.: American Geological Institute.

Averitt, P., 1975. *Coal Resources of the United States, January 1, 1974.* U.S. Geological Survey, Bulletin 1412; Washington D.C.: U.S. Department of Interior; 131 pp.

BLS (Bureau of Labor Statistics), 2006. *Current Population Survey.* Washington, D.C.: U.S. Department of Labor. Available online at http://www.bls.gov/cps/; accessed March 2007.

BNSF (BNSF Railway Company), 2005. Form 10-K for the fiscal year ended December 31, 2005, filed with the U.S. Securities and Exchange Commission. Available online at http://www.bnsf.com/investors/secfilings/10K_railway_2005.pdf; accessed March 2007.

BP (British Petroleum), 2006. *BP Statistical Review of World Energy 2006.* Available online at http://www.bp.com/multipleimagesection.do?categoryId=6840&contentId=7021557; accessed March 2007.

Bradshaw, J., S. Bachu, D. Bonijoly, R. Burruss, N.P. Christensen, and O.M. Mathiassen, 2005. *Draft Discussion Paper from the Task Force for Reviewing and Identifying Standards with Regards to CO_2 Storage Capacity Measurement (Version 2).* Carbon Sequestration Leadership Forum, CLSF-T-2005-9; 16 pp. Available online at http://www.cslforum.org/documents/Taskforce_Storage_Capacity_Estimation_Version_2.pdf; accessed December 2006.

Bradshaw, J., S. Bachu, D. Bonijoly, R. Burruss, S. Holloway, N.P. Christensen, and O.M. Mathiassen, 2006. *CO_2 storage capacity estimation: Issues and development of standards.* Paper presented at the 8th International Conference on Greenhouse Gas Control Technology. Trondheim, Norway, June 19-22; 5 pp.

Brennan, S.T., and R.C. Burruss, 2006. *Specific Sequestration Volumes: A Useful Tool for CO_2 Storage Capacity Assessment.* U.S. Geological Survey, Open-File Report 03-452; 12 pp. Available online at http://pubs.usgs.gov/of/2003/of03-452/of03-452-tagged.pdf; accessed April 2007.

Bullinger, C., M. Ness, and N. Sarunac, 2006. *Coal creek prototype fluidized bed coal dryer: performance improvement, emissions reduction, and operating experience.* Presentation, 31st International Technical Conference on Coal Utilization and Fuel Systems, Clearwater, Fla., May 21-25. Available online at http://www.netl.doe.gov/technologies/coalpower/cctc/ccpi/pubs/GRE/GREClr06_1.pdf; accessed March 2007.

Burke, F., and D. Carter, 2006. *CURC/EPRI Technology Roadmap Update.* Washington, D.C.: Coal Utilization Research Council (CURC) and Electric Power Research Institute (EPRI), September 20. Available online at http://www.coal.org/PDFs/jointroadmap2006.pdf; accessed April 2007.

Carter, M.D., T.J. Rohrbacher, D.D. Teeters, D.C. Scott, L.M. Osmonson, G.A. Weisenfluh, E.I. Loud, R.S. Sites, A.G. Axon, M.E. Wolfe, and L.J. Lentz, 2001. Coal Availability, Recoverability, and Economic Evaluations of Coal Resources in the Northern and Central Appalachian Basin Coal Regions. Chapter J *in: Northern and Central Appalachian Basin Coal Regions Assessment Team: 2000 Resource Assessment of Selected Coal Beds and Zones in the Northern and Central Appalachian Basin Coal Regions.* Denver, Colo., U.S. Geological Survey Professional Paper 1625-C, pp. J1-J43. Available online at http://pubs.usgs.gov/prof/p1625c/; accessed June 2007.

CAST (Center for Advanced Separation Technologies), 2003. *CAST Technology Roadmap.* Available online at http://www.cast.org.vt.edu/Mission/roadmap2003.pdf; accessed April 2007.

DOE (Department of Energy), 2004. *Program Facts: Clean Coal Power Initiative (CCPI).* Available online at http://www.fossil.energy.gov/programs/powersystems/cleancoal/ccpi/Prog052_4P.pdf; accessed March 2007.

DOE, 2005. *The Value of Economic Dispatch. A Report to Congress Pursuant to Section 1234 of the Energy Policy Act of 2005, November 7, 2005.* Available online at http://www.oe.energy.gov/DocumentsandMedia/value.pdf; accessed February 2007.

DOE, 2006. *National Electric Transmission Congestion Study, August 2006.* Available online at http://www.oe.energy.gov/DocumentsandMedia/Congestion_Study_2006-9MB.pdf; accessed March 2007.

DOE, 2007a. *Technologies: Carbon Sequestration.* Available online at http://www.netl.doe.gov/technologies/carbon_seq/index.html; accessed February 2007.

DOE, 2007b. Draft National Interest Electric Transmission Corridor Designations. Office of Electricity Delivery and Energy Reliability. *Federal Register,* 72 (87) (May 7): 25837-25923.

DOE, 2007c. *Carbon Sequestration Atlas of the United States and Canada.* Available online at http://www.netl.doe.gov/publications/carbon_seq/atlas/ATLAS.pdf; accessed June 2007.

DOE/NRC (Department of Energy/Natural Resources Canada), 2004. *Final Report on the August 14, 2003 Blackout in the United States and Canada: Causes and Recommendations.* U.S.-Canada Power System Outage Task Force; 22 pp.

EC (European Commission), 2003. *World Energy, Technology and Climate Policy Outlook 2030— WETO.* EUR 20366; 137 pp. Available online at http://ec.europa.eu/research/energy/pdf/weto_final_report.pdf; accessed April 2007.

EIA (Energy Information Administration), 1999. *U.S. Coal Reserves: 1997 Update.* Available online at http://tonto.eia.doe.gov/FTPROOT/coal/052997.pdf; accessed March 2007.

EIA, 2005a. *Coal News and Markets, Week of May 15, 2005.* Available online at http://tonto.eia.doe.gov/FTPROOT/coal/newsmarket/coalmar050515.html; accessed February 2007.

EIA, 2005b. *Coal News and Markets, Week of May 22, 2005.* Available online at http://tonto.eia.doe.gov/FTPROOT/coal/newsmarket/coalmar050522.html; accessed February 2007.

EIA, 2005c. *Coal News and Markets, Week of November 6, 2005.* Available online at http://tonto.eia.doe.gov/FTPROOT/coal/newsmarket/coalmar051106.html; accessed February 2007.

EIA, 2005d. *Coal Production Data File, 2005.* Available online at http://www.eia.doe.gov/cneaf/coal/page/coalpublic05.xls; accessed March 2007.

EIA, 2006a. *Annual Energy Review 2005.* Report No. DOE/EIA-0384(2005). Washington, D.C.: U.S. Department of Energy. Available online at http://www.eia.doe.gov/emeu/aer/contents.html; accessed April 2007.

EIA, 2006b. *Annual Coal Report 2005.* DOE/EIA-0584. Washington, D.C.: U.S. Department of Energy. Available online at http://www.eia.doe.gov/cneaf/coal/page/acr/acr_sum.html; accessed March 2007.

EIA, 2006c. *U.S. Coal Consumption by End-Use Sector: Data for Second Quarter, 2006.* Available online at http://www.eia.doe.gov/cneaf/coal/quarterly/html/t28p01p1.html; accessed December 2006.

EIA, 2006d. *Annual Energy Outlook 2006 with Projections to 2030.* DOE/EIA-0383(2006). Washington, D.C.: U.S. Department of Energy. Available online at http://www.eia.doe.gov/oiaf/archive/aeo06/pdf/0383(2006).pdf; accessed April 2007.

EIA, 2006e. *Energy Market Impacts of Alternative Greenhouse Gas Intensity Reduction Goals.* SR/OIAF/2006-01. Washington, D.C.: U.S. Department of Energy. Available online at http://www.eia.doe.gov/oiaf/servicerpt/agg/pdf/sroiaf(2006)01.pdf; accessed June 2007.

EIA, 2006f. *International Energy Outlook.* DOE/EIA-0484(2006). Washington, D.C.: U.S. Department of Energy. Available online at http://www.eia.doe.gov/oiaf/archive/ieo06/pdf/0484(2006).pdf; accessed June 2007.

EIA, 2006g. *Domestic Distribution of U.S. Coal by Origin State, Consumer, Destination and Method of Transportation, 2004.* Washington, D.C.: U.S. Department of Energy. Available online at http://www.eia.doe.gov/cneaf/coal/page/coaldistrib/2004/d_04state.pdf; accessed February 2007.

EIA, 2006h. *Average Utility Contract Coal Transportation Rate per Ton-Mile by Transportation Mode, 1979-1997.* Washington, D.C.: U.S. Department of Energy. Available online at http://www.eia.doe.gov/cneaf/coal/ctrdb/tab37.html; accessed February 2007.

EIA, 2007a. *Annual Energy Outlook Retrospective Review: Evaluation of Projections in Past Editions (1982-2006).* Washington, D.C.: U.S. Department of Energy. Available online at http://www.eia.doe.gov/oiaf/analysispaper/retrospective/pdf/0640(2006).pdf; accessed June 2007.

EIA, 2007b. *Glossary: Longwall Mining.* Available online at http://www.eia.doe.gov/glossary/glossary_l.htm; accessed April 2007.

English, G., 2006. *In the Matter of Discussions with Utility and Railroad Representatives on Market and Reliability Matters.* Testimony before the Federal Energy Regulatory Commission, Docket No. AD06-8-000, June 6.

EPA (U.S. Environmental Protection Agency), 1997. *The Benefits and Costs of the Clean Air Act, 1970 to 1990.* Available online at http://yosemite.epa.gov/ee/epa/eermfile.nsf/vwAN/EE-0295-1.pdf/$File/EE-0295-1.pdf/; accessed May 2006.

EPA, 1998. *National Air Quality and Emissions Trends Report, 1998.* Available online at http://www.epa.gov/air/airtrends/aqtrnd98/; accessed March 2007.

EPA, 2005. Rule To Reduce Interstate Transport of Fine Particulate Matter and Ozone (Clean Air Interstate Rule); Revisions to Acid Rain Program; Revisions to the NO_X SIP Call. *Federal Register,* 70 (91) (May 12): 25261-25310.

EPA, 2006a. *Fact Sheet: Final Revisions to the National Ambient Air Quality Standards for Particle Pollution (Particulate Matter).* Available online at http://www.epa.gov/oar/particlepollution/pdfs/20060921_factsheet.pdf; accessed September 2006.

EPA, 2006b. *Inventory of U.S. Greenhouse Gas Emissions and Sinks: 1990-2004.* Available online at http://www.epa.gov/climatechange/emissions/downloads06/06_Complete_Report.pdf; accessed February 2007.

EPA, 2006c. National Ambient Air Quality Standards for Particulate Matter; Final Rule. *Federal Register,* 71 (200) (October 17): 61143-61233.

EPA, 2007. *Air Quality and Emissions—Progress Continues in 2006.* Available online at http://www.epa.gov/airtrends/econ-emissions.html; accessed May 2007.

EPRI (Electric Power Research Institute), 2006a. *Competitiveness of Coal—What Is Being Planned?* Presentation by Stu Dalton, Gulf Coast Power Association, April 6. Available online at http://www.gulfcoastpower.org/default/s06-dalton.pdf; accessed June 2007.

EPRI, 2006b. *Feasibility Study for an Integrated Gasification Combined Cycle Facility at a Texas Site.* Report No.1014510, Palo Alto, Calif.: EPRI, October. Available online at http://www.epriweb.com/public/000000000001014510.pdf; accessed June 2007.

ExxonMobil, 2005. *The Outlook for Energy: A View to 2030.* Available online at http://library.corporate-ir.net/library/11/115/115024/items/176641/2005%20Energy%20Outlook-WB-12-9-05.pdf; accessed March 2007.

Feeley, T.J., III, 2006. *U.S. DOE's Hg Control Technology R&D Program Review.* Presentation, 2006 Mercury Control Technology Conference, December 11-13. Available online at http://www.netl.doe.gov/publications/proceedings/06/mercury/presentations/Feeley_Overview-Presentation_121106.pdf; accessed March 2007.

Fiscor, S., 2005. Prep plant population rebounds: U.S. Prep Plant Census 2005. *Coal Age* (October): 20-26, 30-31.

FRA (Federal Railroad Administration), 2004. *Impact of the Staggers Rail Act of 1980.* Available online at http://www.fra.dot.gov/downloads/policy/staggers_rail_act_impact.pdf; accessed March 2007.

FRA, 2006. Statement of the Honorable Joseph H. Boardman, Federal Railroad Administrator before the U.S. House of Representatives, Committee on Transportation and Infrastructure, Subcommittee on Railroads, April 26. Available online at http://www.fra.dot.gov/downloads/pubaffairs/capacity_final.pdf; accessed February 2007.

Frailey, S.M., R.J. Finley, and T.S. Hickman, 2006. CO_2 sequestration: Storage capacity guideline needed. *Oil and Gas Journal Online,* 104(30); 8 pp. Available online at http://www.ogj.com/currentissue/index.cfm?p=7&v=104&i=30; accessed June 2007.

Freme, F., 2006. *U.S. Coal Supply and Demand: 2005 Review.* Washington, D.C., U.S. Energy Information Administration; 14 pp. Available online at http://www.eia.doe.gov/cneaf/coal/page/special/feature05.pdf; accessed June 2007.

Gale, J., and J. Davidson, 2004. Transmission of CO_2—Safety and economic considerations. *Energy,* 29: 1319-1328.

Hamberger, E.R., 2006. In the Matter of Discussions with Utility and Railroad Representatives on Market and Reliability Matters, Testimony the Federal Energy Regulatory Commission, Docket No. AD06-8-000, June 15. Available online at http://www.aar.org/PubCommon/Documents/Testimony/AAR%20FERC%20Final.pdf; accessed March 2007.

IEA (International Energy Agency), 2004. *World Energy Outlook, 2004.* Paris: International Energy Agency; 577 pp.

IEA, 2006a. *World Energy Outlook, 2006.* Paris: International Energy Agency; 600 pp.

IEA, 2006b. *Energy Technology Perspectives: Scenarios & Strategies to 2050.* Paris: International Energy Agency; 484 pp.

IEAGHG (International Energy Agency Greenhouse Gas R&D Programme), 2006. *Estimating Future Trends in the Cost of CO_2 Capture Technologies.* Report No. 2006/5. Prepared by Carnegie Mellon University, Cheltenham, U.K.

IPCC (Intergovernmental Panel on Climate Change), 2001. *Climate Change 2001: The Scientific Basis. Contribution of Working Group I to the Third Assessment Report of the Intergovernmental Panel on Climate Change.* J.T. Houghton, Y. Ding, D.J. Griggs, M. Noguer, P.J. van der Linden, X. Dai, K. Maskell, and C.A. Johnson (eds.). Cambridge, U.K.: Cambridge University Press; 881 pp.

IPCC, 2005. *Carbon Dioxide Capture and Storage.* B. Metz, O. Davidson, H.C. de Coninck, M. Loos, and L.A. Meyer (eds.). IPCC Special Report. New York: Cambridge University Press; 442 pp.

IPCC, 2007. Summary for Policymakers. *In: Climate Change 2007: The Physical Science Basis. Contribution of Working Group I to the Fourth Assessment Report of the Intergovernmental Panel on Climate Change.* S. Solomon, D. Qin, M. Manning, Z. Chen, M. Marquis, K.B. Averyt, M.Tignor, and H.L. Miller (eds.). Cambridge, U.K.: Cambridge University Press.

IWUB (Inland Waterway Users Board), 2005. *IWUB 19th Annual Report to the Secretary of the Army and the United States Congress.* Available online at http://www.iwub.iwr.usace.army.mil/ IWUBReport2005.pdf; accessed March 2007.

KGS (Kentucky Geological Survey), 2006. *Methods of Mining.* Available online at http://www.uky. edu/kgs/coal/coal_mining.htm, accessed February 2007.

Knoy, M., 2006. Shaping a better future for waterways interests. Presentation to the Inland Waterways Navigational Conference, Memphis, Tenn., March 14-16.

Kowalski, C.A., 2005. *KFx Clean Coal Technology.* Presentation to Indiana Department of Environmental Management. Available online at http://www.in.gov/idem/programs/air/workgroups/ mercury/feb05/idem_final.ppt#276; accessed March 2007.

LCA (Lake Carriers' Association), 2006. *Press release: Great Lakes Coal Trade Down 2.1 percent in 2006.* Available online at http://www.lcaships.com/coal1206-text.pdf; accessed March 2007.

McGinty, K.A., 2004. *Waste coal incentives.* Testimony before the Pennsylvania Senate Environmental Resources and Energy Committee, September 8. Available online at http://www.depweb. state.pa.us/dep/cwp/view.asp?a=3&q=474575; accessed March 2007.

McLennan, R., 2006. Testimony before the U.S. Senate, Committee on Energy and Natural Resources, May 25. Available online at http://energy.senate.gov/public/index.cfm?FuseAction=Hearings. Testimony&Hearing_ID=1560&Witness_ID=4407; accessed June 2007.

Merritt, R.D., and C.C. Hawley, 1986. *Map of Alaska's Coal Resources.* Fairbanks: Alaska Division of Geological and Geophysical Surveys, Special Report 37. MIOF (Mining Industry of the Future), 2000. *Mineral Processing Technology Roadmap*, Washington, D.C.: U.S. Department of Energy, Office of Energy Efficiency and Renewable Energy; 20 pp.

Mohl, W., 2006. In the Matter of Discussions with Utility and Railroad Representatives on Market and Reliability Matters. Testimony before the Federal Energy Regulatory Commission, Docket No. AD06-8-000, June 6.

MSTTC (Mine Safety Technology and Training Commission), 2006. *Improving Mine Safety Technology and Training: Establishing U.S. Global Leadership.* Available online at http://www. coalminingsafety.org/documents/msttc_report.pdf; accessed June 2007.

NARUC (National Association of Regulatory Utility Commissioners), 2006. *Resolution Urging Legal and Regulatory Reform to Improve Railroad Shipper Rates and Quality of Service*, adopted February 16. Available online at http://www.naruc.org/associations/1773/files/EL-3Reformto ImproveRailroadShipperRates.pdf; accessed June 2007.

NAS (National Academy of Sciences), 2005. *Joint Science Academies' Statement: Global Response to Climate Change.* Statement of science academies of Brazil, Canada, China, France, Germany, India, Italy, Japan, Russia, United Kingdom, and the United States, June 7. Available online at http://nationalacademies.org/onpi/06072005.pdf; accessed December 2006.

NCC, 2006. *Coal: America's Energy Future, Volumes I and II.* Washington, D.C.: National Coal Council; 132 pp and 105 pp.

NCEP (National Commission on Energy Policy), 2004. *Ending the Energy Stalemate: A Bipartisan Strategy to Meet America's Energy Challenges.* Washington, D.C.: National Commission on Energy Policy; 128 pp.

NESCAUM (Northeast States for Coordinated Air Use Management), 2005. *Assessment of Control Technology Options for BART-Eligible Sources: Steam Electric Boilers, Industrial Boilers, Cement Plants and Paper and Pulp Facilities.* Available online at http://www.nescaum.org/ documents/bart-control-assessment.pdf; accessed December 2006.

NETL (National Energy Technology Laboratory), 2006a. *Key Issues and Mandates.* Available online at http://www.netl.doe.gov/KeyIssues/index.html; accessed February 2006.

NETL, 2006b. *International Carbon Capture and Storage Projects Overcoming Legal Barriers*. Report No. DOE/NETL-2006/1236. Prepared by Science Applications International Corporation for U.S. Department of Energy, Morgantown, W. Va., National Energy Technology Laboratory, June 23.

NETL, 2007. *Tracking New Coal-fired Power Plants: Coal's Resurgence in Electric Power Generation*. Available online at http://www.netl.doe.gov/coal/refshelf/ncp.pdf; accessed March 2007.

NMA (National Mining Association), 1998. *The Future Begins with Mining*. Washington, D.C.: National Mining Association; 17 pp.

NMA, 2006a. *2005 Coal Producer Survey*. Available online at http://www.nma.org/pdf/coal_producer_survey2005.pdf; accessed March 2007.

NMA, 2006b. *Trends in U.S. Coal Mining, 1923-2005*. Available online at http://www.nma.org/pdf/c_trends_mining.pdf; accessed February 2007.

NRC (National Research Council), 1975. *Underground Disposal of Coal Mine Wastes*. Washington, D.C.: National Academy Press; 172 pp.

NRC, 1981. *Disposal of Excess Spoil from Coal Mining and the Surface Mining Control and Reclamation Act of 1977*. Washington, D.C.: National Academy Press; 207 pp.

NRC, 1994. *Drilling and Excavation Technologies for the Future*. Washington, D.C.: National Academy Press; 176 pp.

NRC, 1995. *Coal: Energy for the Future*. Washington, D.C.: National Academy Press; 304 pp.

NRC, 1996. *Mineral Resources and Sustainability: Challenges for Earth Scientists*. Washington, D.C.: National Academy Press; 26 pp.

NRC, 1999. *Meeting U.S. Energy Resource Needs: The Energy Resources Program of the U.S. Geological Survey*. Washington, D.C.: National Academy Press; 80 pp.

NRC, 2000. *Vision 21: Fossil Fuel Options for the Future*. Washington, D.C.: National Academy Press; 158 pp.

NRC, 2001. *Climate Change Science: An Analysis of Some Key Questions*. Washington, D.C.: National Academy Press; 42 pp.

NRC, 2002a. *Coal Waste Impoundments: Risks, Responses, and Alternatives*. Washington, D.C.: National Academy Press; 244 pp.

NRC, 2002b. *Evolutionary and Revolutionary Technologies for Mining*. Washington, D.C.: National Academy Press; 102 pp.

NRC, 2002c. *Estimating the Public Health Benefits of Proposed Air Pollution Regulations*. Washington, D.C.: The National Academies Press; 170 pp.

NRC, 2003a. *Review of DOE's Vision 21 Research and Development Program—Phase 1*. Washington, D.C.: The National Academies Press; 108 pp.

NRC, 2003b. *The Carbon Dioxide Dilemma: Promising Technologies and Policies*. Washington, D.C.: The National Academies Press; 150 pp.

NRC, 2004a. *The Hydrogen Economy: Opportunities, Costs, Barriers, and R&D Needs*. Washington, D.C.: The National Academies Press; 256 pp.

NRC, 2004b. *The Engineer of 2020: Visions of Engineering in the New Century*. Washington, D.C.: The National Academies Press; 118 pp.

NRC, 2005a. *Interim Report of the Committee on Changes in New Source Review Programs for Stationary Sources of Air Pollutants*. Washington, D.C.: The National Academies Press; 242 pp.

NRC, 2005b. *Educating the Engineer of 2020: Adapting Engineering Education to the New Century*. Washington, D.C.: The National Academies Press; 208 pp.

NRC, 2006. *Managing Coal Combustion Residues in Mines*. Washington, D.C.: The National Academies Press; 256 pp.

NRC, 2007a. *Rising Above the Gathering Storm: Energizing and Employing America for a Brighter Economic Future*. Washington, D.C.: The National Academies Press; 590 pp.

NRC, 2007b. *Mining Safety and Health Research at NIOSH*. Washington, D.C.: The National Academies Press; 228 pp.

NTIS (National Technical Information Service), 2007. *Bureau of Mines Publications and Journal Articles 1910-1996.* Available online at http://www.ntis.gov/products/specialty/bom.asp?loc=4-5-1; accessed May 2007.

NYC (New York City Energy Policy Task Force), 2004. *New York City Energy Policy: An Electricity Resource Roadmap.* Available online at http://www.nyc.gov/html/om/pdf/energy_task_force. pdf; accessed June 2007.

PDEP (Pennsylvania Department of Environmental Protection), 2002. *Report of the Governor's Commission on Quecreek Inundation.* Available online at http://www.dep.state.pa.us/hosting/ minesafetycommission/; accessed February 2007.

Peterson, D.J., T. LaTourrette, and J.T. Bartis, 2001. *New Forces at Work in Mining: Industrial Views of Critical Technologies.* Santa Monica, Calif.: Rand Corporation, Science and Technology Policy Institute; 92 pp.

PJM (PJM Interconnection, LLC), 2007. PJM 2006 Regional Transmission Expansion Plan, version 1, February 27. Available online at http://www.pjm.com/planning/reg-trans-exp-plan.html; accessed June 2007.

Rubin, E.S., 2007. An Overview of CO_2 Capture Technology for Fossil Fuel Power Plants. Presentation to EPA Advanced Coal Technology Work Group. Washington, D.C.: U.S. Environmental Protection Agency. Available online at http://www.epa.gov/air/caaac/coaltech/2007_02_rubin. pdf; accessed March 2007.

Rubin, E.S., S. Yeh, M. Antes, M. Berkenpas, and J. Davison, 2007. Use of experience curves to estimate the future cost of power plants with CO_2 capture. *International Journal of Greenhouse Gas Control,* 1(2): 188-197.

STAPPA/ALAPCO (State and Territorial Air Pollution Program Administrators and Association of Local Air Pollution Control Officials), 2005. *Regulating Mercury from Power Plants: A Model Rule for States and Localities.* Available online at http://www.4cleanair.org/FinalMercury ModelRule-111405.pdf; accessed December 2006.

TVA (Tennessee Valley Authority), 2005. *Final Environmental Assessment, Gallatin Fossil Plant Rail Coal Unloading and Blending Facility.* Available online at http://www.tva.gov/environment/reports/gallatin/fea.pdf; accessed March 2007.

UPC (Union Pacific Corporation), 2006. Form 10-K for the fiscal year ended December 31, 2005, filed with the U.S. Securities and Exchange Commission, February 24, 2006. Available online at http://www.up.com/investors/attachments/secfiling/2006/upc10k_022406.pdf; accessed March 2007.

UPC/BNSF (Union Pacific Corporation and BNSF Railway Company), 2006. *UP, BNSF Announce Southern Powder River Basin Joint Line $100 Million Capacity Expansion Plan.* Press release, May 8. Available online at http://www.uprr.com/newsinfo/releases/capital_ investment/2006/0508_sprb.shtml; accessed March 2007.

USACE (U.S. Army Corps of Engineers), 2006. *Waterborne Commerce of the United States, Calendar Year 2004. Part 5—National Summaries.* Available online at http://www.iwr.usace.army. mil/ndc/wcsc/pdf/wcusnatl04.pdf; accessed December 2006.

USBM (U.S. Bureau of Mines), 1975a. *The Reserve Base of U.S. Coals by Sulfur Content, Part I: The Eastern States.* Information Circular 8680. Washington D.C.: U.S. Department of the Interior.

USBM, 1975b. *The Reserve Base of U.S. Coals by Sulfur Content, Part II: The Western States.* Information Circular 8693. Washington D.C.: U.S. Department of the Interior.

USDOI (U.S. Department of the Interior), 1976. *Coal Resource Classification System of the U.S. Bureau of Mines and the U.S. Geological Survey.* Geological Survey Bulletin 1450-B. Washington D.C.: U.S. Department of the Interior. Available online at http://pubs.usgs.gov/bul/b1450b/ b1450.htm; accessed June 2007.

Watzmann, B., 2004. *The Aging of the Energy and Minerals Workforce: A Crisis in the Making?* Statement before the U.S. House of Representatives, Committee on Resources, Subcommittee on Energy and Mineral Resources. Available online at http://www.nma.org/pdf/cong_test/ watzman_070804.pdf; accessed February 2007.

WEC (World Energy Council), 2004. *2004 Survey of Energy Resources,* 20th ed. Amsterdam: Elsevier; 464 pp.

WEC, 2006. *Energy Data Centre, Global Energy Scenarios to 2050 and Beyond.* Available online at http://www.worldenergy.org/wec-geis/edc/scenario.asp; accessed December 2006.

Wilkes, D., 2006. *Coal-Based Generation Reliability.* Testimony before U.S. Senate, Committee on Energy and Natural Resources, May 25. Available online at http://energy.senate.gov/public/ index.cfm?FuseAction=Hearings.Testimony&Hearing_ID=1560&Witness_ID=4410; accessed June 2007.

Winschel, R.A., 1990. The relationship of carbon dioxide emissions with coal rank and sulfur content. *Journal of the Air and Waste Management Association*, 40: 861-865.

Wise, M., J. Dooley, R. Dahowski, and C. Davidson, 2007. Modeling the impacts of climate policy on the deployment of carbon dioxide capture and geologic storage across electric power regions in the United States. *International Journal of Greenhouse Gas Control*, 1(2): 261-270.

Wood, G.H., Jr., T.M. Kehn, M.D. Carter, and W.C. Culbertson, 1983. *Coal Resource Classification System of the U.S. Geological Survey.* U.S. Geological Survey Circular 891, Washington D.C.: U.S. Department of the Interior; 65 pp.

Appendixes

Appendix A

Committee and Staff Biographies

Corale L. Brierley, Chair (NAE), provides technical and business consultation to the mining and chemical industries and government organizations through Brierley Consultancy LLC. Previously, Dr. Brierley worked as chemical microbiologist at New Mexico Institute of Mining and Technology, as chief of environmental process development for Newmont Mining Corporation, as general partner at VistaTech Partnership, Ltd., and as president of Advanced Mineral Technologies, Inc. Her research interests include the treatment and management of metal-bearing aqueous, solid, and radioactive wastes and biotechnology applied to mine production. She is on the International Advisory Committee for the Biohydrometallurgy Symposia and the Editorial Board for *Hydrometallurgy Journal*. Dr. Brierley served on the National Academy of Engineering's 2007 Nominating Committee, Committee on Membership, and the Grainger Challenge Prize Committee. She has served on several National Research Council (NRC) committees, including the Committee on the Superfund Site Assessment and Remediation in the Coeur d'Alene River Basin, the Committee on Technology for the Mining Industries, the Committee on Earth Resources, the Committee on Novel Approaches to the Management of Greenhouse Gases, and chair of the Committee to Review the USGS (U.S. Geological Survey) Mineral Resources Program. She also chairs the Engineering Panel for the Ford Foundation Diversity Fellowships Program. Dr. Brierley holds a Ph.D. in environmental sciences from the University of Texas at Dallas.

Francis P. Burke retired at the end of 2006 after serving in the research and development department of CONSOL Energy Inc. (and its predecessor organizations) since 1975. In 1996 he became vice president of Research and Development, with general management responsibility for CONSOL's research program. The goal of CONSOL's R&D program is to identify, develop, and apply technology that advances the near-term and strategic interests of CONSOL's coal, natural gas, and other business units. In 2004, he became vice president of science & technology, with responsibilities in the areas of energy and environmental policy, and since retiring he continues to consult for CONSOL Energy on these issues. He is a member of the Advisory Board of the University of Kentucky Center for Applied Energy Research and serves on the NRC's Committee on Earth Resources, the Advisory Board of the Pittsburgh Coal Conference, the Advisory Board of the Dominion Center for Engineering and the Environment at the University of Pittsburgh, and the Department of Energy's (DOE's) working group on Strategic Initiatives for Coal and Power. Dr. Burke is the author of more than 80 scientific papers and publications, and holds five U.S. patents on coal-related technology. He is a member of Phi Kappa Phi and the American Chemical Society, was twice the recipient of the American Chemical Society Fuel Chemistry Division's R. A. Glenn Award, and received the Senator Jennings Randolph Lifetime Achievement Award presented by the Washington Coal Club in December 2006. Dr. Burke holds a Ph.D. in physical chemistry from Iowa State University, and he has completed the Executive Program at the Darden School of the University of Virginia.

James C. Cobb is an adjunct professor in the Department of Geological Sciences at the University of Kentucky, the director of the Kentucky Geological Survey, and the state geologist of Kentucky. Dr. Cobb has been with the Kentucky Geological Survey for the past 27 years. He has served in the capacity of a geologist, a section head, and an assistant state geologist for research. His research interests include coal geology with respect to coal availability and resources of Kentucky; estimating compliance coal resources for Kentucky; deposition, resources, sulfur, mining; basin evolution with respect to mineral formation in coal; hydrogeology with respect to groundwater aquifers in North Africa; modern analogues of coal formation in Indonesia; and industrial minerals with respect to Cretaceous-Tertiary gravel, and Pleistocene sand and gravel in Illinois. Throughout his career, Dr. Cobb has published in more than 60 journals, survey publications, special papers, abstracts, and reports. Dr. Cobb received his Ph.D. from the University of Illinois, Urbana.

Robert B. Finkelman was formerly a senior scientist and project chief for the Eastern Energy Resources Team at the U.S. Geological Survey. His research interests include coal chemistry and medical geology. Dr. Finkelman has a diverse professional background—he worked at the USGS for 32 years, at Exxon for 7 years, and has experience as a consultant and as a college instructor. Most of Dr. Finkelman's professional career has been devoted to understanding the properties of coal and how these properties affect coal's technological performance, economic by-product potential, and environmental and health impacts. For the past 10 years he has devoted his efforts to developing the field of medical geology. Dr. Finkelman is a fellow of the Geological Society of America (GSA), is the author of more than 550 publications, and is a recipient of the Gordon H. Wood Jr. Memorial Award from the American Association of Petroleum Geologists Eastern Section and the Cady Award from the GSA's Coal Geology Division. Dr. Finkelman received his Ph.D. in chemistry from the University of Maryland.

William Fulkerson is presently a senior fellow with the Institute for a Secure and Sustainable Environment (ISSE) at the University of Tennessee. Prior to his retirement in 1994 from the Oak Ridge National Laboratory, he was associate laboratory director for energy and environmental technologies. His current interests include global sustainability issues with emphasis on energy and environmental technologies and policies. Since 1994, he has chaired the DOE Laboratory Energy R&D Working Group (LERDWG), an organization of energy R&D managers from 14 DOE labs including all of the national labs concerned with energy R&D. During 1999 and 2000, LERDWG helped the under secretary of energy analyze the DOE energy R&D portfolio with respect to its adequacy for making progress on DOE strategic goals related to the environment, the economy, and national security. More recently, LERDWG has assisted DOE in the planning of the National Climate Change Technology Initiative of the Bush Administration and with drafting a strategic plan for the Clean Energy Technology Export Initiative. Dr. Fulkerson was a member of the Energy R&D Panel of the President's Committee of Advisors on Science and Technology, and he chaired the panel's task force on fossil energy. He was a member of the NRC Board on Energy and Environmental Systems from 1996 to 2002. Dr. Fulkerson received his Ph.D. in chemical engineering from Rice University.

Harold J. Gluskoter is a scientist emeritus with the U.S. Geological Survey. His research interests include national and international coal resource assessments. Dr. Gluskoter is one of the nation's leading coal geologists and he played a significant role in the national coal assessment. He was awarded the Geological Society of America's Gilbert H. Cady Award for contributions that advance the field of coal geology in North America. His research interests, in addition to coal resource assessments, have included coal geochemistry as it is related to coal utilization and the environment and more recent studies of the potential for sequestering carbon dioxide in coal beds. Dr. Gluskoter also brings a state agency perspective through his former service with the Illinois State Geological Survey. Dr. Gluskoter received his Ph.D. in geology from the University of California, Berkeley.

Michael E. Karmis is the Stonie Barker Professor of Mining and Minerals Engineering and director of the Virginia Center for Coal and Energy Research at

Virginia Polytechnic Institute. His broad research interests are in mine planning and design, ground control, carbon sequestration, and the sustainable development of energy and mineral resources. An author of more than 150 publications, Dr. Karmis has been active in consulting with the minerals industry, consulting companies, government organizations, and legal firms. He served as the 2002 president of the Society for Mining, Metallurgy and Exploration (SME) and the 2002-2003 President of the Society of Mining Professors. He is a distinguished member of the SME, a fellow of the Institute of Quarrying, and a fellow of the Institute of Materials, Minerals and Mining. Dr. Karmis received his Ph.D. from the University of Strathclyde, U.K.

Klaus S. Lackner is the Ewing-Worzel Professor of Geophysics in the Department of Earth and Environmental Engineering at Columbia University. He previously held postdoctoral positions at the California Institute of Technology and the Stanford Linear Accelerator Center before joining Los Alamos National Laboratory in 1983 in the Theoretical Division. He also held several management positions, including acting associate laboratory director for strategic and supporting research. Currently, he is developing innovative approaches to energy issues of the future. He has been instrumental in forming ZECA, the Zero Emission Coal Alliance, which was an early industry-led effort to develop coal power with zero emissions to the atmosphere. His most recent work is on environmentally acceptable technologies for the use of fossil fuels. Dr. Lackner received his Ph.D. in theoretical particle physics from Heidelberg University, Germany.

Reginald E. Mitchell is an associate professor in the Department of Mechanical Engineering at Stanford University. He is the current director of the High Temperature Gasdynamics Laboratory, a research laboratory within the Thermosciences Group that houses research in combustion science, pollution science, fluid mechanics, spray dynamics, plasma science, materials synthesis, and laser-based optical diagnostics. Dr. Mitchell's research interests include coal and biomass combustion and gasification, pyrite combustion, pollutant formation and destruction during combustion, and hydrocarbon flame chemistry and structure. He is an active member of the Combustion Institute, having held several positions on the Executive Committee of its Westerns States Section and is a member of the National Organization of Black Chemists and Chemical Engineers, having served as chair of its Western Region for several years. Dr. Mitchell holds a Sc.D. from the Massachusetts Institute of Technology.

Raja V. Ramani (NAE) is emeritus George H. Jr. and Anne B. Deike Chair in Mining Engineering and emeritus professor of mining and geoenvironmental engineering at the Pennsylvania State University. Dr. Ramani holds M.S. and Ph.D. degrees in mining engineering from Penn State where he has been on the faculty since 1970. His research activities include mine health, safety, productivity, environment, and management; flow mechanisms of air, gas, and dust in mining environs; and innovative mining methods. Dr. Ramani has been a consultant to the United Nations, World Bank, and National Safety Council and has received numerous awards from academia and technical and professional societies. He was the 1995 president of the Society for Mining, Metallurgy, and Exploration, Inc. He served on the U.S. Department of Health and Human Service's (DHHS) Mine Health Research Advisory Committee (1991-1998). He has served on a number of NRC committees, including the Committee on Coal Waste Impoundments, the Panel on Technologies for the Mining Industries, the Committee on the Review of NIOSH (National Institute of Occupational Safety and Health) Research Programs, and was chair of the Committee to Review the NIOSH Mining Safety and Health Research Program. In 2002, he chaired the Pennsylvania Governor's Commission on Abandoned Mine Voids and Mine Safety that was set up immediately following the Quecreek Mine inundation incident and rescue.

Jean-Michel M. Rendu (NAE) is a mining consultant and retired vice president for resources and mine planning at Newmont Mining Corporation. Dr. Rendu was previously an associate with Golder Associates in Denver, Colorado; an adjunct professor at the Colorado School of Mines; an assistant professor of mining engineering at the University of Wisconsin, Madison; and head of operations research with Anglovaal in Johannesburg, South Africa. Dr. Rendu is recognized as a leader in the development of national and international standards for the evaluation and public reporting of mineral resources and reserves. His current interests are in optimizing the evaluation, development, and operation of mining projects using appropriate mathematical and managerial technology; as well as drilling and sampling methods, deposit modeling, mine design, ore control, reconciliation of production results with exploration models, and development of computerized systems that facilitate and speed up data collection, quality control, data analysis, and decision making. Dr. Rendu received his doctor of engineering science from Columbia University.

Edward S. Rubin is a professor in the Department of Engineering and Public Policy and the Department of Mechanical Engineering at Carnegie Mellon University. He holds a chair as the Alumni Professor of Environmental Engineering and Science and was founding director of the university's Center for Energy and Environmental Studies and the Environmental Institute. His teaching and research are in the areas of energy utilization, environmental control, technology innovation, and technology-policy interactions, with a particular focus on issues related to coal utilization and global climate change. He is a fellow member of the American Society of Mechanical Engineers, a past chairman of its Environmental Control Division, and recipient of the Air and Waste Management Association Lyman A. Ripperton Award for distinguished achievements as an educator. He has served on advisory committees to the U.S. Department of Energy, the U.S. Environmental Protection Agency (EPA), the Intergovernmental Panel on Climate Change, and the National Academies, including two terms on the Board on Energy and Environmental Studies. Dr. Rubin received his Ph.D. in mechanical engineering from Stanford University.

Samuel A. Wolfe is chief counsel for the New Jersey Board of Public Utilities, where he leads a team working on federal and regional energy policy issues and oversees a staff of legal specialists working on state regulatory matters involving natural gas, electricity, water, and telecommunications and cable television. Previously, as assistant commissioner at the New Jersey Department of Environmental Protection (NJDEP), he supervised NJDEP's Division of Air Quality, Division of Water Quality, and Division of Environmental Safety and Health. He led NJDEP's efforts to reduce mercury emissions from New Jersey's coal-fired power plants and other sources, helped develop key aspects of a seven-state agreement to cap greenhouse gas emissions from power plants, and worked to strengthen federal regulation of power plant emissions. Mr. Wolfe has also served as environmental policy manager for PSEG Services Corporation, where he led the company's environmental due diligence for potential acquisitions of electric generating facilities; made proposals to the EPA, environmental groups, and other stakeholders to reform the New Source Review program under the Clean Air Act; and worked to resolve environmental permitting and enforcement issues with regulatory agencies. Mr. Wolfe holds a B.A. from Cornell University and a J.D. from the University of North Carolina at Chapel Hill.

National Research Council Staff

David A. Feary is a senior program officer with the NRC's Board on Earth Sciences and Resources (BESR) and staff director of BESRs Committee on Seismology and Geodynamics. Prior to joining the NRC, he spent 15 years as a research scientist with the marine program at the Australian Geological Survey Organisation (now Geoscience Australia). During this time, he participated in numerous national and international research cruises to better understand the role of climate as a primary control on carbonate reef formation and to improve understanding of cool-water carbonate depositional processes and controls. Dr. Feary received his Ph.D. from the Australian National University.

Appendix B

Presentations to the Committee

The committee received the following presentations at public committee meetings and mine tours, held during January-November, 2006:

Mike Adamczyk, Joy Mining Machinery—*Coal Mining Equipment Trends and Future Equipment Supply Issues.*

Carl O. Bauer, DOE-National Energy Technology Laboratory—*DOE's Coal Technology Development Mission and Expectations for NRC Coal Study.*

Peter J. Bethell, Arch Coal, Inc.—*Coal Preparation, Current Status and the Way Ahead.*

Perry Bissell, John T. Boyd Co.—*U.S. Coal Markets through 2020: CoalVision™ 2006.*

Paul P. Bollinger, Jr., U.S. Air Force, Office for Installations, Environment and Logistics—*Meeting the Mission for Domestic Alternative Fuels.*

Richard Bonskowski, DOE-Energy Information Administration—*The Demonstrated Reserve Base of Coal–Status and History.*

Joe Cerenzia, CONSOL Energy, Inc.—*CONSOL Mine 84.*

Gregory E. Conrad, Interstate Mining Compact Commission—*State Perspective on Coal Research and Development Needs in the Area of Environmental Protection.*

John Craynon, DOI-Office of Surface Mining—*An Overview of the Office of Surface Mining and OSM's Perspective on the Coal Research, Technology and Resource Assessments to Inform Energy Policy Study.*

Rob Donovan, U.S. Energy Association—*Global Coal Usage and Reserves.*

Tom Dower, Senator Specter's Office—*Study Background and Congressional Expectations.*

Nick Fedorko, West Virginia Geological and Economic Survey—*A Report on the State of Knowledge of Coal Resources and Reserves in West Virginia.*

Ari Geertsema, University of Kentucky—*Coal to Liquid Fuels in the U.S.: Research and Technology.*

Steve Gigliotti, DOL-Mine Safety and Health Administration—*Overview of Technical Support: U.S. Department of Labor, Mine Safety and Health Administration.*

R. Güner Gürtunca, DHHS-National Institute for Occupational Safety and Health—*Coal Mine Safety and Health Research at Pittsburgh Laboratory of NIOSH.*

David Hawkins, Natural Resources Defense Council—*Speeding CCS Deployment.*

Peter Holman, Caterpillar Global Mining—*Global/USA Energy.*

Connie Holmes, National Mining Association—*Perspective of the National Mining Association.*

Mike Hood, CRC Mining—*Technologies That Will Be Needed for Mining in 2025* and *The CRCMining Academic-Industry Research Model.*

James R. Katzer, Massachusetts Institute of Technology—*Advanced Coal-Based Power Generation Technology.*

Larry Kellerman, Goldman Sachs & Co—*Overcoming the Not-So-Hidden Barriers to the Expansion of Coal Generation.*

Julianne M. Klara, DOE-National Energy Technology Laboratory—*Benefits of Clean Coal R&D Program.*

Mo Klefeker, Black Hills Corporation—*Energy for the West: Wire Versus Rail Decisions.*

Jeffrey L. Kohler, DHHS-National Institute for Occupational Safety and Health—*NIOSH's Mission and Expectations for the NRC Coal Study.*

John Langton, DOL-Mine Safety and Health Administration—*Overview of Coal Mine Safety and Health.*

John A. Lewis, DOI-Bureau of Land Management—*National Science Academy Presentation by the Bureau of Land Management.*

Alexander Livnat, U.S. Environmental Protection Agency—*EPA's Role in Regulation and R&D of Coal-Related Activities.*

James Luppens, U.S. Geological Survey—*Overview of Coal Research Activities at the U.S. Geological Survey.*

Gerald H. Luttrell, Virginia Tech,—*Status and Needs of the Coal Preparation Industry.*

John Moran, Jones Walker, Waterways Council, Inc.—*The National Public Policy Organization Advocating a Modern and Well-Maintained System of Ports and Inland Waterways.*

M. Granger Morgan, Carnegie Mellon University—*Security and Reliability of the Electric Power Transmission and Distribution System.*

Mike Mosser, DOE-National Energy Technology Laboratory—*Mining Industry of the Future.*

John Novak, Electric Power Research Institute—*Coal RD&D Insights from CoalFleet for Tomorrow®.*

Karen Obenshain, Edison Electric Institute—*Coal Rail Transportation: The Electric Power Industry's View.*

Bruce Peterson, Oak Ridge National Laboratory—*Forecasting the Transportation Environment for Coal.*

Brenda S. Pierce, U.S. Geological Survey—*The USGS Energy Resources Program: An Overview.*

Jacek Podkanski, International Energy Agency—*Perspective of the International Energy Agency.*

Craig Rockey, American Association of Railroads—*Railroads and Coal.*

Timothy Rohrbacher, U.S. Geological Survey—*Resource and Reserve: USGS Coal Resource and Reserve Assessments.*

Scott Sitzer, DOE-Energy Information Administration—*Coal Activities of the Energy Information Administration.*

Neil Stiber, U.S. Environmental Protection Agency—*EPA's Role in Regulation and R&D of Coal-Related Activities.*

Eugene Trisko, United Mine Workers of America—*UMWA Perspectives on Energy from Coal.*

Joe Vacarri, Rio Tinto Energy America—*RTEA-Cordero Rojo Mine.*

Ted Venners, KFx, Inc.—*Overview of KFx K-Fuel Coal Processing.*

Kimery Vories, DOI-Office of Surface Mining—*Environmental Issues with the Potential to Impact Coal Production over the Next 25 Years.*

Franz Wuerfmannsdobler, Senator Byrd's Office—*Study Background and Congressional Expectations.*

Ben Yamagata, Coal Utilization Research Council—*CURC Perspective on Coal Research, Technology, and Resource Assessments to Inform Energy Policy Study.*

Appendix C

Federal Support for Coal Research

T his appendix presents a brief overview of federal agency coal-related research programs and legislation and the current levels of federal agency support for coal-related research and technology development throughout the coal life cycle, including funding trends over the last 5 to 10 years. Unless otherwise noted, all funding data are presented in terms of 2005 dollars.[1]

FEDERAL AGENCY COAL-RELATED RESEARCH PROGRAMS AND LEGISLATION

One of the earliest federal agencies to become involved in mining research was the U.S. Geological Survey (USGS), which, under congressional authorization, commenced a minerals information collection activity in 1882. However, the history of federally funded mining research in the United States is to a large extent tied to the creation of the U.S. Bureau of Mines (USBM) in the Department of the Interior by the Organic Act of 1910. As a result of subsequent congressional actions, the role of USBM extended to health and safety in mines, testing of fuels, and technical processes of production and use. Enactment of the Leasing Act in 1920 resulted in USBM becoming responsible for supervising mining operations on public lands. The Coal Mine Inspection Act of 1941 authorized USBM to enter and inspect mines and recommend corrective actions. Although portions of the USBM were transferred to other agencies at various times, the principal focus for mining research, related to health, safety, and productivity, remained with USBM. The USBM has been credited with speeding up the introduction of many health

---

[1] All figures are adjusted using the inflation calculator at http://data.bls.gov/cgi-bin/cpicalc.pl.

and safety practices, such as generalized rock dusting, permissible explosives, explosion and fire control measures, improved mine rescue procedures, methane control and drainage, noise control, human factors, and electrical safety.

In 1960, the Office of Coal Research was separated from USBM and charged to develop new and more efficient methods of mining, processing, and utilizing coal. The program was transferred to the Energy Research and Development Administration (ERDA) in 1974, and it eventually became incorporated in the Department of Energy (DOE) where today it forms part of the Office of Fossil Energy.

The impact of the Federal Coal Mine Health and Safety Act of 1969 on coal mine design and operations continues to this day. It mandated health and safety standards for coal mines and directed USBM to conduct the research necessary to eliminate coal mine health and safety hazards. This legislation also directed that mining health research be conducted in the Department of Health, Education, and Welfare. Following closure of the USBM in 1995 and the eventual transfer of health and safety research to the National Institute for Occupational Safety and Health (NIOSH) in 1997, the NIOSH Mining Program became the principal focus for mining health and safety research. The Occupational Safety and Health Act of 1970 created the National Institute of Occupational Safety and Health to focus on health research. The mining industry regulatory functions of the USBM were separated from its mining research functions in 1973. Under the 1977 Mine Health and Safety Act, these functions were entirely transferred to the Mine Safety and Health Administration (MSHA) in the Department of Labor.

With passage of the Surface Mining Control and Reclamation Act of 1977 (SMCRA), the Office of Surface Mining Reclamation and Enforcement (OSM) was created in the Department of the Interior. SMCRA specified the planning and design requirements for mining from both health and safety and environmental perspectives. It also required mining companies to submit plans and designs for approval and provided agencies with the power to monitor and inspect mines for compliance purposes. Finally, it established research and training centers in the states dealing with various aspects of mineral production. The centers were very active in OSM's first decade, supporting mineral education and creating state mining research institutes.

CURRENT FEDERAL AGENCIES SUPPORTING COAL-RELATED RESEARCH AND TECHNOLOGY DEVELOPMENT

More than $538 million was allocated by U.S. government agencies for coal-related research and technology development in 2005 (see Table 7.2). For this report, funding estimates were compiled through an interactive process between the committee and agency staff. First, the committee requested budgets for coal research and development (R&D), and based on responses from the agencies, the committee chose to include activities that were variously described as pure research, applied science, pilot-scale testing, technical support, applied engineering projects, and demonstration projects. Funding estimates were requested for the 1995, 2000, and 2005 fiscal years, so that funding trends might be discerned. The data were analyzed by the committee, and the committee's interpretations of the data were sent back to agency staff for confirmation before being included in this analysis.

Department of Energy (DOE)

The **Energy Information Administration**[2] **(EIA)**, created in 1977, is a policy-independent statistical agency within DOE. It provides data, forecasts, and analyses to support energy policy and public understanding of energy trends. By mandate, EIA neither formulates nor advocates any particular policy conclusions. EIA tracks coal prices, production, reserves, distribution, consumption, stocks, imports, and exports nationally and internationally, and issues a broad range of weekly, monthly, and annual reports on energy production, stocks, demand, imports, exports, and prices. It also prepares analyses and special reports on topics of current interest that are widely used by federal and state agencies, industry, media, researchers, consumers, and educators. In 2005, EIA received only 0.07 percent of overall federal coal R&D funding. EIA's coal-related budget, in constant 2005 dollars, decreased by 77 percent between 1995 and 2005.

The **Office of Fossil Energy**[3] **(FE)** has the primary mission to ensure that the country continues to rely on clean, affordable energy from traditional fuel resources, including coal. FE is responsible for implementing two major coal-related research programs, the 10-year, $2 billion Clean Coal Power Initiative to develop a new generation of environmentally benign clean coal technologies, and the FutureGen Initiative, a $950 million coal-fueled prototype plant that will co-produce electricity and hydrogen while minimizing the release of air pollutants and greenhouse gases into the atmosphere. Other coal R&D programs include pollution control innovations for traditional power plants (including mercury reduction), improved gasification technologies, advanced combustion systems, development of stationary power fuel cells, improved turbines for future coal-based combined cycle plants, and the creation of a portfolio of technologies that can capture and permanently store greenhouse gases. Most FE coal-related research is administered by the **National Energy Technology Laboratory**[4] **(NETL)**, which has a focus on creating commercially viable solutions to national energy and environmental problems. In addition to research conducted on-site, NETL's project portfolio includes R&D conducted through partnerships, cooperative research and development agreements, financial assistance, and contractual

---

[2]See http://www.eia.doe.gov/neic/aboutEIA/aboutus.html.
[3]See http://www.fossil.energy.gov.
[4]See http://www.netl.doe.gov/about/index.html.

arrangements with other national laboratories, academic institutions, and the private sector. Of all the federal agencies carrying out coal R&D in 2005, DOE-FE (including NETL) had by far the largest budget, 82 percent of the total. FE research predominantly addresses coal utilization issues, with an allocation for carbon sequestration research that has increased rapidly over the past six years. The FE budget for coal R&D, in constant 2005 dollars, increased by almost 25 percent between 1995 and 2005.

The **Office of Energy Efficiency and Renewable Energy**[5] **(DOE-EERE)**, through its Industrial Technologies Program, funded the Mining Industry of the Future initiative to support engineering and technology development designed to improve the energy efficiency, resource utilization, and competitiveness of the mining industry. Although this program was not focused solely on coal mining and processing, many of the program outputs were applicable to these phases of the coal fuel cycle. Program budgets increased from 1999 to 2003 but have decreased since 2004, and new funding for this program has now been terminated as the program is closed out. For this analysis, only the coal-specific elements have been included and the cross-cutting components have not been considered. In 2005, this program made up less than 2 percent of federal coal R&D. Of the coal-related projects undertaken during the eight-year history of the program, 87 percent addressed mining and processing issues and 13 percent responded to safety and health issues; funding for 2005 was distributed according to this ratio.

The **Office of Electricity Delivery and Energy Reliability**[6] **(DOE-OE)** was created in 2003 to "lead national efforts to modernize the electric grid, enhance security and reliability of the energy infrastructure and facilitate recovery from interruptions in energy supply." The two major R&D programs funded by OE are high-temperature superconductivity, focused on developing pre-commercial prototypes of electric power equipment, and transmission reliability, focused on deployment of real-time monitoring capabilities. Because it is not possible to divide OE R&D budgets into coal-related and other electricity delivery components, the OE R&D appropriations were divided according to the proportion (53 percent) of national electricity generation supplied by coal-fired power plants. On this basis, OE supported 9 percent of federal coal-related R&D funding in 2005.

Department of Health and Human Services (DHHS)

The **National Institute for Occupational Safety and Health**[7] was established by the Occupational Safety and Health Act of 1970 (the OSH Act) to

---

[5]See http://www.eere.energy.gov/.

[6]See http://www.oe.energy.gov/.

[7]See http://www.cdc.gov/niosh/about.html.

prevent work-related illness, injury, disability, and death by gathering information, conducting scientific research, and translating the knowledge gained into products and services. Focusing on high-risk sectors such as mining, NIOSH performs basic research and field research studies on worker safety and health, develops recommendations for occupational safety and health standards, conducts on-site investigations to determine the toxicity of materials used in workplaces, identifies engineering controls for existing equipment, conducts training and employee education, and funds research by other agencies or private organizations through grants, contracts, and other arrangements (NRC, 2007b). The Federal Mine Safety and Health Amendments Act of 1977 delegated additional authority to NIOSH for coal mine health research, including the development of recommendations for mine health standards to be regulated by MSHA, administering a medical surveillance program for miners, conducting on-site investigations in mines similar to those authorized for general industry under the OSH Act, and testing and certifying personal protective equipment and hazard measurement instruments. In 2005, NIOSH research support for coal mining safety and health amounted to approximately 4 percent of all federal coal R&D. In constant 2005 dollars, the NIOSH coal R&D budget has increased slightly (11 percent) over the past five years. However, when compared with the USBM coal mine safety and health program, the NIOSH coal R&D funding in 2005 was only 42 percent of USBM funding in 1994 (the last full year of USBM funding).

Department of the Interior (DOI)

The **Bureau of Land Management**[8] **(BLM)** is responsible for resource management and conservation for vast areas of public lands, primarily in the western United States. BLM sells leases for a significant proportion of the nation's coal mines, with royalty and bonus bid income of more than $1.2 billion in FY 2005. BLM does not undertake coal-related R&D activities itself, but works with other agencies (e.g., OSM, USGS) to address issues that require R&D.

The **Office of Surface Mining Reclamation and Enforcement**[9] **(OSM)** has a primarily regulatory role stipulated by the Surface Mining Control and Reclamation Act. OSM's main coal-related objectives are to ensure that coal mines are operated in a manner that protects citizens and the environment to ensure that the land is restored to beneficial use following mining, and to mitigate the effects of past mining by pursuing reclamation of abandoned mines. This is accomplished by providing direct technical assistance to deal with specific mining and reclamation problems, maintaining automated systems, providing tools, and transferring technical capabilities through training, consultation, forums, and conferences. In the past, OSM addressed problems related to reclamation projects and regulatory

---

[8]See http://www.blm.gov/wo/st/en/info/About_BLM.2.html.
[9]See http://www.osmre.gov/mission.html.

implementation through cooperative research efforts with other bureaus. OSM now undertakes a small amount of coal-related research itself, focused primarily on reclamation, in support of its regulatory role. There was a significant increase in OSM's coal R&D budget between 1995 and 2005, by more than 130 percent in constant 2005 dollars, but this research still comprised only 0.1 percent of the total federal coal R&D budget in 2005. Since 2005, OSM's R&D funding has increased even more dramatically—from $0.6 million in 2005 to $1.4 million in 2006—primarily to provide increased support for the applied science and underground mine mapping programs.

United States Geological Survey[10] **(USGS)** coal research activities, within its Energy Resources Program, focus primarily on assessments of resources and reserves. Additional efforts focus on compilation of coal quality information and research on the environmental and human health impacts of coal extraction and combustion. The USGS coal program accounts for almost 2 percent of total federal coal R&D funding. Between 1995 and 2005, the USGS coal R&D budget gradually decreased by 29 percent as inflation eroded essentially flat budget allocations.

Department of Labor (DOL)

The **Mine Safety and Health Administration**[11] **(MSHA)** administers the provisions of the Federal Mine Safety and Health Act of 1977 (Mine Act) to enforce compliance with mandatory safety and health standards as a means to eliminate fatal accidents, reduce the frequency and severity of nonfatal accidents, minimize health hazards, and promote improved safety and health conditions in the nation's mines. MSHA provides technical support and training services to its personnel and to personnel from the mining industry through its Pittsburgh Safety and Health Technology Center and the National Mine Health and Safety Academy. Because it is primarily a regulatory agency, MSHA's involvement in coal mine research is mostly as a "customer" for NIOSH research activities. However, it does undertake field investigations, laboratory studies, and cooperative research activities related to health and safety issues, and evaluates new equipment and materials for use in mines. MSHA also supports state miner training activities through its states-grant program, and it works collaboratively through partnerships and coordinated research to ensure that mining technology, practices, and controls are developed and implemented to protect miner health or safety. The committee estimated that 5 percent of the MSHA technical support funding, almost $1.3 million in 2006, could be considered coal mining safety and health research. This amounted to almost 0.25 percent of the total federal coal R&D for

---

[10]See http://energy.usgs.gov/coal.html.
[11]See http://www.msha.gov/MSHAINFO/MISSION.HTM.

that year. MSHA's coal-related R&D budget, in constant 2005 dollars, decreased by almost 13 percent between 1995 and 2006.

U.S. Environmental Protection Agency (EPA)

The **Office of Research and Development**[12] **(EPA-ORD)** supports limited research, both internally and through extramural funding, to support the EPA's primarily regulatory role to implement federal laws designed to protect human health and the environment. The environmental problems associated with active and abandoned mines, particularly land reclamation, water quality maintenance, and the proper handling and disposal of the spoils and wastes from mining operations (e.g., mountain top coal mining, coal combustion residues), are the focus of EPA attention. EPA funds almost 2 percent of the total federal coal R&D, with most research focused on utilization issues (e.g., mercury and other emissions). EPA's coal-related R&D budget, in constant 2005 dollars, remained approximately constant between 1995 and 2006.

National Science Foundation (NSF)

The **National Science Foundation**[13] **(NSF)** provides funding for fundamental research across all areas of science and engineering. Funding for coal-related research is distributed across NSF directorates and program areas; consequently there was no single contact point for information concerning coal-related activities. The committee determined NSF's budget by searching NSF's online listing of grants awarded between 1995 and 2005 for references to coal. Approximately 525 grants were reviewed as a result of the online search, and 30 were identified as falling within the purview of the committee. These were primarily in the areas of coal utilization (64 percent) and mining and processing (21 percent). Because almost all were multiyear grants, the total annual award amount was calculated assuming uniform funding throughout the life of each grant. Consequently, it was not feasible to determine trend data. NSF's coal-related awards make up a little more than 0.5 percent of the total federal coal R&D budget.

---

[12]See http://www.epa.gov/ord/htm/aboutord.htm.
[13]See http://www.nsf.gov/about/glance.jsp.

Appendix D

Data Tables: U.S. and World Coal Reserves

The most widely referenced data tables that report the coal reserves of the United States and the rest of the world are presented below. The data reported for the United States are the Demonstrated Reserve Base (DRB) and the Estimated Recoverable Reserves (ERR) (see Table D.1); these are reported annually by the Energy Information Administration (EIA) and are described in Box 3.1.

The most recent World Energy Council (WEC) compilation of worldwide coal reserves from 75 countries—the twentieth edition of *Survey of Energy Resources*—was published in 2004 (see Table D.2). The WEC provides the following definitions to the nation member committees, which then provide the data that are the bases for Table D.2 (WEC, 2004):

• **Proved amount in place** is the resource remaining in known deposits that has been carefully measured and assessed as exploitable[1] under present and expected local economic conditions with existing available technology; and

• **Proved recoverable reserves** are the tonnage *within* the proved amount in place category that can be recovered in the future under present and expected local economic conditions with existing available technology.

The WEC must rely on data submitted by each country, and "it is well recognized that each country tends to have its own notion of what constitutes resources and reserves" (WEC, 2004, p. xi.).

---

[1]Equivalent to "minable"; wording is from WEC (2004).

TABLE D.1 U.S. Coal Reserve Data by State for 2005, ERR and DRB by Mining Method for 2005 (million short tons)

State	Underground Minable Coal		Surface Minable Coal		Total	
	ERR	DRB	ERR	DRB	ERR	DRB
Alabama	508	1,007	2,278	3,198	2,785	4,205
Alaska	2,335	5,423	499	687	2,834	6,110
Arizona	—	—	—	—	NA	NA
Arkansas	127	272	101	144	228	417
Colorado	6,015	11,461	3,747	4,762	9,761	16,223
Georgia	1	2	1	2	2	4
Idaho	2	160	—	—	2	160
Illinois	27,927	87,919	10,073	16,550	38,000	104,469
Indiana	3,620	8,741	434	742	4,054	9,483
Iowa	807	1,732	320	457	1,127	2,189
Kansas	—	—	681	972	681	972
Kentucky total	7,411	17,055	7,483	12,965	14,894	30,020
Eastern	658	1,178	5,214	9,337	5,872	10,516
Western	6,753	15,877	2,269	3,628	9,022	19,504
Louisiana	—	—	312	422	312	422
Maryland	317	578	44	65	361	643
Michigan	55	123	3	5	59	128
Mississippi	—	—	—	—	NA	NA
Missouri	689	1,479	3,157	4,510	3,847	5,989
Montana	35,922	70,958	39,021	48,272	74,944	119,230
New Mexico	2,801	6,156	4,188	5,975	6,988	12,131
North Carolina	5	11	—	—	5	11
North Dakota	—	—	6,906	9,053	6,906	9,053
Ohio	7,719	17,546	3,767	5,754	11,486	23,300
Oklahoma	574	1,231	226	323	800	1,554
Oregon	6	15	2	3	9	17
Pennsylvania, total	10,710	23,221	1,044	4,251	11,754	27,472
Anthracite	340	3,844	420	3,355	760	7,198
Bituminous	10,370	19,377	625	896	10,994	20,274
South Dakota	—	—	277	366	277	366
Tennessee	280	510	179	264	459	774
Texas	—	—	9,534	12,385	9,534	12,385
Utah	2,514	5,128	212	268	2,726	5,396
Virginia	2,949	1,130	171	562	3,121	1,693
Washington	1,030	1,332	6	8	1,036	1,340
West Virginia	15,576	29,184	2,382	3,775	17,958	32,960
Wyoming	22,950	42,500	17,657	21,319	40,607	63,819
U.S. Total	**152,850**	**334,876**	**114,705**	**158,059**	**267,554**	**492,935**

NOTES:

NA = This estimated value is not available due to insufficient or inadequate data or model performance.

The DRB and ERR as of January 1, 2006, incorporate revisions made to eliminate a discrepancy between data expressed by coal rank versus data allocated to British thermal unit (Btu) ranges. The minor differences resulted from the fact that coal rank classifications are based in part, but not entirely, on Btu content. EIA's data—originally allocated to Btu ranges for coal supply and demand modeling—had been used to approximate the ERR by coal rank in the early 1990s. Over time, the small differences between resources and reserves by coal rank and by Btu ranges became significant due to cumulative depletion adjustments. The January 1, 2006, data include internal additions to coal tonnages by Btu ranges to identify the coal ranks where more than one rank occurs in borderline resource areas and to unify the tonnage totals. Recoverable coal reserves at producing mines represent the quantity of coal that can be recovered (i.e., mined) from existing coal reserves at reporting mines.

EIA's ERR include the coal in the DRB considered recoverable after excluding coal estimated to be unavailable due to land use restrictions or currently economically unattractive for mining after applying assumed mining recovery rates.

The effective date for the DRB, as customarily worded, is "Remaining as of January 1, 2006." These data are contemporaneous with the RRPM, customarily presented as of the end of the past year's mining—in this case, December 31, 2005. Current or recent mining in a state does not imply those data for a DRB and ERR.

The DRB includes publicly available data on coal mapped to measured and indicated degrees of accuracy and found at depths and in coalbed thicknesses considered technologically minable at the time of determinations.

All reserve expressions exclude silt, culm, refuse bank, slurry dam, and dredge operations. RRPM excludes mines producing less than 10,000 short tons, which are not required to provide reserves data.

SOURCES: EIA Form EIA-7A, *Coal Production Report*; MSHA, Form 7000-2, *Quarterly Mine Employment and Coal Production Report*; and EIA estimates.

TABLE D.2 Proved International Recoverable Coal Reserves at End of 2002 (million tonnes)

Country	Bituminous (including anthracite)	Subbituminous	Lignite	Total
Algeria	40			40
Botswana	40			40
Central African Republic			3	3
Congo (Democratic Republic)	88			88
Egypt (Arab Republic)	21			21
Malawi		2		2
Morocco	N[a]			N
Mozambique	212			212
Niger	70			70
Nigeria	21	169		190
South Africa	48,750			48,750
Swaziland	208			208
Tanzania	200			200
Zambia	10			10
Zimbabwe	502			502
Total Africa	**50,162**	**171**	**3**	**50,336**
Canada	3,471	871	2,236	6,578
Greenland		183		183
Mexico	860	300	51	1,211
United States of America	111,338	101,978	33,327	246,643
Total North America	**115,669**	**103,332**	**35,614**	**254,615**
Argentina		424		424
Bolivia	1			1
Brazil		10,113		10,113
Chile	31	1,150		1,181
Colombia	6,230	381		6,611
Ecuador				24
Peru	960		24	1,060
Venezuela	479		100	479
Total South America	**7,701**	**12,068**	**124**	**19,893**
Afghanistan	66			66
China	62,200	33,700	18,600	114,500
India	90,085		2,360	92,445
Indonesia	740	1,322	2,906	4,968
Japan	259			359
Kazakhstan	28,151		3,128	31,279
Korea (DPR)	300	300		600
Korea (Republic)		80		80
Kyrgyzstan			812	812
Malaysia	4			4
Mongolia[b]				
Myanmar	2			2
Nepal		1		1
Pakistan		60	2,990	3,050

TABLE D.2 (continued)

Country	Bituminous (including anthracite)	Subbituminous	Lignite	Total
Philippines	22	144	70	236
Taiwan, China	1			1
Thailand			1,354	1,354
Turkey	278	761	3,147	4,186
Uzbekistan	1,000		3,000	4,000
Vietnam	150			150
Total Asia	**183,358**	**36,368**	**38,367**	**258,093**
Albania			794	794
Austria			20	20
Bulgaria	4	91	2,092	2,187
Croatia	6		33	39
Czech Republic	2,094	3,242	216	5,552
France	15			15
Germany	183		6,556	6,739
Greece			3,900	3,900
Hungary	198	199	2,960	3,357
Ireland	14			14
Italy		27	7	34
Netherlands	497			497
Norway		5		5
Poland	14,000			14,000
Portugal	3		33	36
Romania	22	3	469	494
Russian Federation	49,088	94,472	10,450	157,010
Serbia and Montenegro	9	656	15,926	16,591
Slovakia	N		172	172
Slovenia		40	235	275
Spain	200	300	30	530
Sweden		1		1
Ukraine	16,274	15,946	1,933	34,153
United Kingdom	220			220
Total Europe	**82,827**	**117,982**	**45,826**	**246,653**
Iran (Islamic Republic)	419			419
Total Middle East	**419**			**419**
Australia	38,600	2,200	37,700	78,500
New Caledonia	2			2
New Zealand	33	205	333	571
Total Oceania	**38,635**	**2,405**	**38,033**	**79,073**
Total World	**478,771**	**272,326**	**157,967**	**909,064**

[a]N represents negligible amounts.
[b]A quantification of proved recoverable reserves for Mongolia is not available.
SOURCE: WEC (2004).

Appendix E

Coal Mining and Processing Methods

This appendix presents additional details on the individual processes that are involved in extraction of coal from surface and underground mines, and the subsequent beneficiation of the coal in coal processing plants to produce a final product.

COAL MINING METHODS

Surface Mining

In surface mining, the ground covering the coal seam (the overburden) is first removed to expose the coal seam for extraction. The elements of a surface mining operation are (1) topsoil removal and storage for later use, (2) drilling and blasting the strata overlying the coal seam, (3) loading and transporting this fragmented overburden material (called spoil), (4) drilling and blasting the coal seam, (5) loading and transporting the coal, (6) backfilling with spoil and grading, (7) spreading top soil over the graded area, (8) establishing vegetation and ensuring control of soil erosion and water quality, and (9) releasing the area for other purposes (Figure E.1). Steep topography, a steeply dipping seam, or multiple seams, all present challenging problems for designing stable slopes and productive operations in surface mining situations.

Surface topography controls which of the surface mining methods—contour mining, area strip mining, or open-pit mining—is employed (see Figure 4.3). These differ principally in the methods employed for loading, transporting, and storing the spoil. Contour mines are common in the hilly Appalachian terrain of the eastern United States where the fragmented overburden has to be transported outside the mining area for placement and storage. In the Midwest, where the surface topography and coal seams are generally flat, it is common to employ area strip mining in which the fragmented overburden is placed directly by large draglines in the space created where coal has been mined (Figure E.1). In some situations in the eastern United States, a coal seam occurring near the top of mountains is exposed by removing the top of the mountain (Figure 4.3) and transporting the fragmented overburden to a nearby valley.

Underground Mining

Underground mining is usually by the room-and-pillar mining or longwall mining method (Figure E.2). Even in mines where the longwall method is the principal extraction method, the development of the mine and the longwall panels is accomplished by room-and-pillar continuous mining. The thickness of the coal seam, the depth and inclination of the coal seam, the nature of roof and floor strata, and the amount of gas contained both in the coal seam and the roof and floor strata are all important for selection of the mining method. Mining difficulties are greatly increased if seams are extremely thick or thin or are steeply inclined. Longwall mining additionally requires large coal reserves to justify the capital cost of longwall equipment.

As surface mining in the Powder River and Rocky Mountain Basins proceeds, it is likely that the stripping ratios (overburden to coal) will exceed an economic limit. If this coal is to be mined at reasonably high recovery rates, it

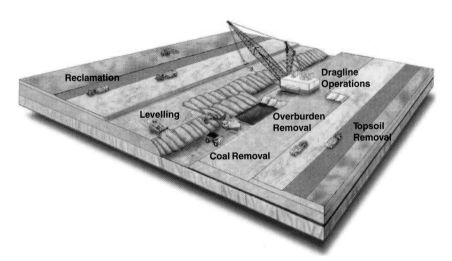

FIGURE E.1 Schematic depiction of the unit operations in a surface coal mine. SOURCE: Royal Utilities.

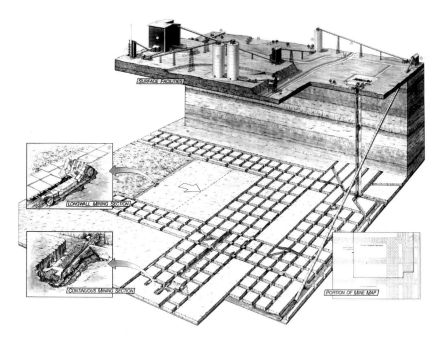

FIGURE E.2 Schematic showing underground coal mine workings. The coal seam is accessed by both a slope and a shaft, shown on the right. The ventilation fan arrangement is shown adjacent to the surface opening of the shaft. The shaft has an elevator for lowering and raising miners and materials. Coal gathered from the workings by various conveyors is transported to the surface by the slope conveyor. The surface features shown are the raw coal storage silo fed by the slope conveyor, the coal preparation plant (the building on the left), the clean coal storage silos in the front, and the train load out. A longwall section and a room-and-pillar continuous miner section are shown. The room-and-pillar section is a five-entry development with rows of four pillars. The longwall face is between two three-entry developments. SOURCE: CONSOL Energy Inc.

will require thick-seam underground mining methods such as large longwalls or multiple slice and/or caving techniques that have not been used in the United States. This will require improvements to mining equipment and practices that are likely to entail research and development (R&D) on mine design, ground control, mine automation, and new systems for protecting worker health and safety.

Room-and-Pillar Mining. In the room-and-pillar method, a set of entries, usually between three and eight, are driven into a block of coal. These entries are connected by cross-cuts, which are usually at right angle to the entries. The entries are commonly spaced from 50 to 100 feet apart, and the cross-cuts are usually about 50 to 150 feet apart. The pillars formed by the entries and cross-cuts may be extracted or left standing depending on mining conditions. In the *conventional room and pillar method*, several pieces of equipment are used in sequence at a working face to extract the coal. These unit operations include the undercutting, drilling, blasting, loading and roof bolting operations. In the *continuous room and pillar method*, the unit operations of undercutting, drilling, and blasting are eliminated and the cutting and loading functions are performed by a mechanical machine—the continuous miner. The room-and-pillar method accounts for 50 percent of the underground production in the United States, and continuous mining makes up more than 90 percent of this production. In both conventional and continuous methods, coal is loaded onto coal transport vehicles and then dumped onto a panel-belt conveyor for transport out of the mine. Once the coal has been cut, the strata above the excavated coal seam are supported by roof bolts. Under favorable conditions, the production from a continuous miner section can exceed 800,000 tons per year per continuous miner.

Longwall Mining. Longwall mining is an automated form of underground coal mining characterized by high recovery and extraction rates, feasible only in relatively flat-lying, thick, and uniform coal beds. A high-powered cutting machine (the shearer) is passed across the exposed face of coal, shearing away broken coal, which is continuously hauled away by a floor-level conveyor system (Figure E.2). Longwall mining extracts all machine-minable coal between the floor and ceiling within a contiguous block of coal, known as a panel, leaving no support pillars within the panel area. Panel dimensions vary over time and with mining conditions but currently average about 900 feet wide (coal face width) and more than 8,000 feet long (the minable extent of the panel, measured in the direction of mining). Longwall mining is done under movable roof supports that are advanced as the bed is cut. The roof in the mined-out area is allowed to fall as the mining advances (EIA, 2007b). The use of longwall mining in underground production has been growing in terms of both amount and percentages, increasing from less than 10 percent of underground production (less than 10 million annual tons) in the late 1960s, to about 50 percent of underground production (more than 200 million annual tons) at present. The production from a longwall mine today (one longwall section and two or three continuous miner sections) can exceed 7 million tons per year. With a second longwall and the necessary complement of continuous miners, production from an underground longwall mine can be well over 10 millions tons per year.

COAL PROCESSING METHODS

The composition of coals mined in different areas can vary widely (Table 4.2). Since the very early days of mining, coal quality has been improved by removing unwanted mineral matter. Over this time, coal preparation plants have evolved considerably, from simple size segregation in the early twentieth century, into lump coal for domestic use and intermediate sizes for industrial use. The fines were rejected as unfit for use, leading to a substantial quantity of coal refuse ("waste coal" or "gob" piles) particularly in the eastern states. The first washing methods were imported from Europe, The "Chance" washer, in which the density differences between coal and mineral matter was exploited to clean raw coal was introduced in 1918. The Chance washer utilized sand and water as a medium. Today, the "heavy-media" process using magnetite is standard for coarse coal cleaning. Attempts to recover middlings and fine coal have continued through the years, and near the middle of the twentieth century, processes to wash and recover fine coal resulted in the introduction of equipment such as centrifuges, froth flotation cells, disc filters, thickeners, cyclones, and thermal dryers.

The unit processes in coal preparation plants vary, but the following sequence of steps is typical.

• *Crushing and breaking*. Run-of-mine coal must be crushed to an acceptable top size for treatment in the preparation plant. Typical crushing and breaking devices are feeder breakers, rotary breakers, hammer mills, and roll crushers.

• *Sizing*. Different cleaning processes are used on different sizes of coal. Therefore raw coal entering the plant will be screened (sieved) into three or four sizes. Clean coal is rarely sized, except for some industrial markets.

• *Storage and stockpiling*. Coal is stored in silos or stockpiled before and after cleaning. Raw coal is stored between the mine and the preparation plant, and clean coal is stored between the preparation plant and product loadout. This is done to provide surge capacity at the interface between the mine and the plant, and between the plant and the loadout, to maintain workable product inventories, and in some cases to control the quality of coal going to a given customer by segregating different products.

• *Density separation*. Raw coal consists of organic and mineral matter components, with specific gravities ranging from 1.30 for the lighter organic material to 2.5 for rock. Coal is cleaned by separating the lower-density organic material from the higher-density refuse. In heavy-media separations, the specific gravity of the medium used for separation, usually a suspension of finely divided magnetite in water, is chosen to achieve a given degree of separation depending on the characteristics of the coal, the desired product quality, and the acceptable level of coal loss to the rejects. In water-only devices such as jigs, spirals, and water-only cyclones, separation is effected by the differential acceleration of coal and mineral particles in water.

• *Froth flotation*. Fine coal particles (i.e., smaller than 0.5 mm) are difficult to separate from mineral matter on a density basis and this fraction usually is cleaned by froth flotation. Froth flotation is a physiochemical process that exploits the selectivity of the attachment of air bubbles to organic coal particle surfaces and in the nonattachment to mineral constituents. Surfactants are used to create a hydrophobic surface on the coal particles to be floated, and a "collector," typically fuel oil, is used to promote agglomeration of the floated particles to facilitate their removal.

• *Coal drying.* Coal preparation plants that employ fine coal cleaning by froth flotation can produce an unacceptable amount of moisture in the product. Thermal drying, in which the wet coal is dried in the hot gas generated by a coal- or gas-fired burner, is used in some plants to reduce the moisture content.

• *Refuse and tailings management.* Waste management is an integral part of coal preparation. Coarse refuse is transported to the solids disposal area, where it can form a tailings impoundment or be placed in a suitable landfill. Tailings (fine solid waste in water) are usually transported by pipeline to an impoundment area where the tailings settle out; the clarified water is reused in the plant.

Coal Preparation Plants

Each year, *Coal Age* magazine conducts a census of coal preparation plants in the United States (Fiscor, 2005). The overall findings of the survey (summarized in Table E.1) are generally accepted within the industry as a reasonably accurate reflection of the condition of the coal preparation industry. According to the *Coal Age* article, "plants reported an average recovery rate of 57%." Given the total raw coal capacity of the surveyed plants (158,187 tons per hour), this corresponds to a clean-coal capacity of 790 million tons per year, assuming 24/7 operation.

The number of preparation plants increased by 53 since the 2000 survey, and at least 10 new plants were built and 25 were significantly upgraded in the

TABLE E.1 Characteristics of Coal Preparation Plants in the United States in 2004, by State

State	Number of Plants	Raw Coal Capacity (ton/hr)	Average Age
West Virginia	66	48,382	24
Kentucky	73	43,320	21
Pennsylvania (bituminous)	20	14,575	30
Virginia	25	10,700	21
Illinois	11	10,450	21
Indiana	19	8,950	17
Alabama	6	8,120	26
Ohio	15	5,360	24
Pennsylvania (anthracite)	15	1,980	35
Maryland	1	1,800	N/A
Colorado	4	1,750	7
Washington	1	1,750	NA
Utah	6	600	NA
Tennessee	2	450	NA
Total	**264**	**158,187**	**23**

SOURCE: Fiscor (2005).

five years to 2005 (Fiscor, 2005). The survey notes that "while they employ new equipment, technology, and circuitry, U.S. prep plants have in general remained essentially the same. The typical U.S. prep plant employs heavy media separation and was built in 1983. It has a raw capacity between 500 and 1,000 tons per hour. Although more plants are employing large diameter cyclones, the plants still rely mainly on heavy-media vessels for primary separation and heavy media cyclones for intermediate separation. For fine coal recovery, the plants prefer spirals. Centrifugal dryers are popular. On the technology side, the industry has not embraced online analysis on a widespread basis, but it has adopted the use of PLCs extensively" (Fiscor, 2005).[1]

---

[1]A PLC is a Programmable Logic Controller.

Appendix F

Acronyms and Abbreviations

ABET	Accreditation Board for Engineering and Technology
ACARP	Australian Coal Association Research Program
ACR	Australian Coal Research
ACT	accelerated technology
AEO	Annual Energy Outlook
AFBC	atmospheric fluidized bed combustion
AML	abandoned mine lands
BCR	Bituminous Coal Research
BLM	Bureau of Land Management
Btu	British thermal unit
CAST	Center for Advanced Separation Technology
CCPI	Clean Coal Power Initiative
CCS	carbon capture and sequestration (*alt.* carbon capture and storage)
CCTDP	Clean Coal Technology Demonstration Program
CP	carbon price
CRCMining	Cooperative Research Centre for Mining
CTL	coal-to-liquids
CURC	Coal Utilization Research Council
DoD	Department of Defense
DOE	Department of Energy
DOE-EERE	Office of Energy Efficiency and Renewable Energy (DOE)
DOE-EIA	Energy Information Administration (DOE)
DOE-FE	Office of Fossil Energy (DOE)
DOE-OE	Office of Electricity Delivery and Energy Reliability (DOE)

DOI	Department of the Interior
DOL	Department of Labor
DRB	Demonstrated Reserve Base
EC	European Commission
EIA	Energy Information Administration (DOE)
EOR	enhanced oil recovery
EPA	U.S. Environmental Protection Agency
EPRI	Electric Power Research Institute
ERDA	Energy Research and Development Administration
ERR	Estimated Recoverable Reserves
FGD	flue gas desulfurization
FP	fuel price
FY	fiscal year
GDP	gross domestic product
GHG	greenhouse gas
GIS	geographic information system
GRE	Great River Energy
Gt	gigaton
GW	gigawatt
IEA	International Energy Agency
IGCC	integrated gasification combined cycle
IOF	Industry of the Future
IPCC	Intergovernmental Panel on Climate Change
IWUB	Inland Waterways Users Board
MSHA	Mine Safety and Health Administration
Mt	megaton
NCEP	National Commission on Energy Policy
NCRA	National Coal Resource Assessment
NEMS	National Energy Modeling System
NETL	National Energy Technology Laboratory
NIOSH	National Institute for Occupational Safety and Health
NMA	National Mining Association
NSF	National Science Foundation
O&G	oil and gas
O&M	operations and maintenance
OECD	Organisation for Economic Co-operation and Development
ORD	Office of Research and Development (EPA)
OSM	Office of Surface Mining Reclamation and Enforcement
PC	pulverized coal
PNNL	Pacific Northwest National Laboratory
PPII	Power Plant Improvement Initiative
PRB	Powder River Basin
R&D	research and development

RD&D	research, development, and demonstration
RRAM	Recoverable Reserves at Active Mines
SMCRA	Surface Mining Control and Reclamation Act
SNG	substitute natural gas (also known as synthetic natural gas)
UIC	Underground Injection Control
UMWA	United Mine Workers of America
UNECE	United Nations Economic Commission for Europe
USACE	U.S. Army Corp of Engineers
USBM	U.S. Bureau of Mines
USGS	U.S. Geological Survey
WEC	World Energy Council
WEO	World Energy Outlook
WETO	World Energy, Technology, and Climate Policy Outlook

Appendix G

Unit Conversion Factors
and Energy Ratings

CONVERSION FACTORS

1 Btu [British thermal unit] = 1.055 kJ [kilojoules] = 252 cal [calories]
1 cal = 0.003967 Btu = 4.184 J
1 quad [quadrillion Btu] = 10^{15} Btu
1 Btu/lb = 0.556 kcal/kg

ENERGY RATINGS

1 tce [tonne of coal equivalent] = 29.308 GJ = 27.778 million Btu
1 toe [tonne of oil equivalent] = 41.868 GJ = 39.683 million Btu
"Coal equivalent" coal = 7,000 kcal/kg
High-rank coal = 7,000 kcal/kg
Low-rank coal = 3,500 kcal/kg
Lignite = 2,700 kcal/kg
1 ton lignite = 0.3 to 0.63 tce (average 0.38)
1 ton subbituminous = 0.78 tce